D1329666

IEE CONTROL ENGINEERING SERIES 12

SERIES EDITORS: G. A. MONTGOMERIE
PROF. H. NICHOLSON

MODELLING
OF DYNAMICAL
SYSTEMS Vol.1

Previous volumes in this series:

3
.1

MODELLING OF DYNAMICAL SYSTEMS Vol.1

Edited by
H. Nicholson, D.Eng., M.A., F.I.E.E., M.I.Mech.E.
Professor of Control Engineering
University of Sheffield
England

PETER PEREGRINUS LTD.
on behalf of the
Institution of Electrical Engineers

Tennessee Tech. Library 329674
Cookeville. Tenn.

Published by: The Institution of Electrical Engineers, London
and New York
Peter Peregrinus Ltd., Stevenage, UK, and New York

© 1980: Institution of Electrical Engineers

All rights reserved. No part of this publication may be reproduced,
stored in a retrieval system or transmitted in any form or by any
means—electronic, mechanical, photocopying, recording or otherwise—
without the prior written permission of the publisher

British Library Cataloguing in Publication Data

Modelling of dynamical systems.
Vol. 1.—(Institution of Electrical Engineers.
IEE control engineering series, vol. 12).
1. System analysis
2. Mathematical models
I. Nicholson, Harold   II. Series
511'.8   QA402   80-40463

ISBN 0-906048-38-9

Typeset by Santype International Limited, Salisbury
Printed in England by A. Wheaton & Co., Ltd., Exeter

Contents

Preface

A significant research effort has been devoted to the mathematical modelling of industrial processes and socio-economic and biological systems, and the work has been published essentially in conference proceedings and learned society journals. Our present aim is to highlight some of this work and to make it readily available in various book volumes as a convenient reference for research workers, postgraduate students and for those in industry and research establishments who are concerned with system modelling and control-system design.

Mathematical modelling has a very important role in systems analysis and design which requires the representation of systems phenomena as a functional dependence between interacting input and output variables. This is illustrated by various case studies, in which the physical processes, particularly, are represented as an interconnection of energy storage and dissipative elements, with the overall model formulated using the basic equations of continuity, momentum, energy and heat transfer and also state relationships. The various subsystem models when linearized, will possess modes of oscillation with a relatively wide range of time constants. Similar modes of behaviour and rhythmic changes affect our social and also biological systems, with high- and low-frequency modes of oscillation associated with different activity levels in a complex interconnected structure. A certain unity of representation thus exists for all dynamical systems and this will become evident through study of the various chapters.

Chapter 1, 'Modelling principles and simulation', introduces the basic methodology for model building based on an assembly of fundamental equations, and also illustrates experimental approaches to model building and parameter estimation. Chapter 2 then demonstrates how insight can be acquired into the fundamental behaviour of physical processes by deriving dynamic parametric models in transfer function matrix form and also multi-variable first-order lag models. The techniques are illustrated with reference to distributed thermodynamic and chemical processes, including a counter-flow liquid/liquid heat exchanger, binary distillation column and tubular chemical reactor.

The modelling of refrigeration and air-conditioning systems is the subject of Chapter 3, and is illustrated by the development of various models including a vegetable quick-freezing tunnel, a laboratory air-conditioning system and a banana-boat hold. Chapter 4 then considers a typical modelling problem for thermal reactor systems, starting, for simplicity, from the one-group neutron diffusion equation. The emphasis is on linear state space models suitable for stability and control studies and the work includes linearization of the diffusion equation with representation of inherent feedback and xenon-induced instability, and the development of low-order lumped models based on a combination of modal and finite-difference methods.

In contrast, Chapter 5 considers the dynamics and control of rigid aerospace vehicles including a guided missile and fixed-wing aircraft. Models of pilot behaviour and of gimballed inertial navigation units incorporating data filtering are also considered. Similar techniques and scale-model tests are then discussed in relation to the modelling of surface ship dynamics in Chapter 6. The use of systems identification techniques, including nonlinear effects in large-vessel surface manoeuvring, and the applications of self-adaptive and self-tuning control strategies for the design of autopilots are also discussed.

The case studies used to illustrate the modelling of dynamical systems are then completed in Chapter 7 with a review of work on biological systems modelling, including eye pupil regulation, respiratory control, gastro-intestinal rhythms, nerve impulses, glucose regulation, and biochemical cell oscillations.

The theme on the modelling of dynamical systems is continued in Volume 2 with contributions on Ironmaking and Steelmaking, Steel Processes, Power, Water and Gas Supply Systems, Coal and Mineral Extraction and Manufacturing Processes.

J.B.

List of contributors

J. B. EDWARDS

Department of Control Engineering, University of Sheffield, Mappin Street, Sheffield
S1 3JD, England

R. W. JAMES

Faculty of Environmental Science & Technology, Polytechnic of the South Bank,
Borough Road, London SE1 OAA

J. R. LEIGH

The Polytechnic of Central London, 115 New Cavendish Street, London W1M 8JS

D. A. LINKENS

Department of Control Engineering, University of Sheffield, Mappin Street, Sheffield
S1 3JD, England

J. M. LIPSCOMBE

Department of Electronic & Control Engineering, Cranfield Institute of Technology,
Cranfield, Bedford MK43 OAL

D. H. OWENS

Department of Control Engineering, University of Sheffield, Mappin Street, Sheffield
S1 3JD, England

Chapter 1

Modelling principles and simulation

J. R. Leigh

1.1 Introduction

Mathematical models can be categorized according to purpose. In the first category are *models to assist plant design and operation*. The second category consists of *models to assist control system design and operation*.

Models to assist plant design and operation

(*a*) Detailed, physically based, often non-dynamic models to assist in fixing plant dimensions and other basic parameters.
(*b*) Economic models allowing the size and product mix of a projected plant to be selected.
(*c*) Economic models to assist decisions on plant modernization.

Models to assist control systems design and operation

(*a*) Fairly complete dynamic models, valid over a wide range of process operations to assist detailed quantitative design of a control system.
(*b*) Simple models based on crude approximation to the plant, but including some economically quantifiable variables, to allow the scope and type of a control system to be decided.
(*c*) Reduced dynamic models for use on-line as part of a control system.

Some types of control systems require an *on-line predictive model*. Such a model introduces special problems in that future values of the uncontrolled plant inputs are not available and estimates must be used. To illustrate this, suppose that a plant using a varying raw material is to be controlled using a predictive model. Assumptions must be made about the composition of the future raw material in order for the predictive model to operate.

Where the model is to be used to assist in the design of an optimal controller a model must be produced that is mathematically tractable. In particular, it will be advantageous if the principal equations are differentiable.

It should be noted that there exist methods for carrying out plant optimization on-line without the use of a model. The performance of the plant is measured and by a succession of controlled perturbations the operating point of the plant is moved to maximize the performance. This approach has been called evolutionary operation. It is applicable to continuous plants whose condition changes only slowly with time.

The *extent of use* of a model will affect its complexity. A model that must represent a range of types of plant and a range of sizes of plant must necessarily be more complex than a model that has to represent a single plant. In particular, the single-plant model can include empirical relations that are valid only for that plant while a further simplification is that plant-dependent coefficients can be regarded as constant.

The information from which a mathematical model is built includes material-flow diagrams, block diagrams, circuit diagrams, algebraic and differential equations and curves and tables. To this must be added the results of experiments and data-logging trials. The resulting mathematical model is assembled into a computer program. The flow diagram for the computer program of a typical mathematical model is shown in Fig. 1.1.

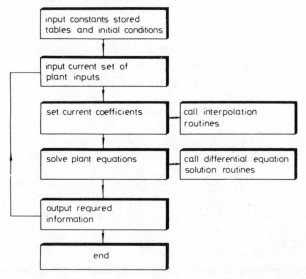

Fig. 1.1   *Flow diagram of the computer program for a typical mathematical model*

1.2 A methodology for model building

The initial stage in model building is to assemble equations representing the physical mechanisms that are believed to be applicable to the plant. (This procedure is described in Section 1.3.) A selection from these equations is then manipulated to obtain a framework for the desired mathematical model.

The different types of information available to the model builder will often be incompatible, because they exist at different theoretical levels, and are impossible to combine. For instance, equations representing the basic process may well be extensive and complex, being based on a rather profound process theory. This gives rise to many internal relations with unknown inaccessible parameters. At the other extreme there will be semi-empirical relations with a small number of variables.

Assuming that the equations can be made compatible, the next stage of equation selection can proceed. All equations that will have significant effect on the model behaviour must obviously be included. Equations whose effect is unknown can also be included at this point—they can be removed later if they have no part to play. The disadvantage of this strategy is that the initial model is likely to be large and complex and it may be difficult in practice to eliminate the redundant equations. The alternative, to produce first a simple model and to refine this until it performs as required, is in practice more attractive for most applications.

Unknown coefficients in the model equations are determined if possible from the process literature or from specially devised tests. They are stored either as numerical values or in the look-up tables. The remaining unknown coefficients in the model must be determined by parameter estimation techniques as described in Section 1.8.

Plant data must be obtained for the purposes of parameter estimation and for model validation. Sufficient data must be available to allow statistically meaningful tests to be carried out and the data must span the range of conditions over which the model has to operate. Obtaining sufficient reliable data of the right type from a large industrial plant is a costly and time-consuming exercise. There is a temptation to use normal operating records but this can be done successfully only in a minority of cases. Usually, a specially conducted data collection trial is required involving augmented instrumentation, controlled plant conditions and specially injected disturbances. Where an existing plant is being modelled, it will usually be the intention to modify the plant, for instance by the application of an improved control system. Thus, the data collection trial must take the plant into regimes that are part of its expected future range of operation rather than its existing, perhaps more limited, range.

For plants not yet in existence, it may be possible to use data from an existing similar plant although even nominally identical plants differ markedly in practice in their numerical coefficients. Difficult coefficients for plants not yet existing will have to be obtained from pilot plant or laboratory tests concentrated on this one aspect. For instance, wind-tunnel tests on a Perspex model may be used to obtain an aerodynamic coefficient. Such an approach can be used only when the model has a sound and well specified theoretical basis.

Once the best coefficient values have been obtained, the model is tested against plant data to determine whether it meets the modelling criterion laid down previously. Except for very simple plants, it is very unlikely that the modelling criterion will be satisfied at the first or second attempt. Model building is essentially iterative as shown in Fig. 1.2.

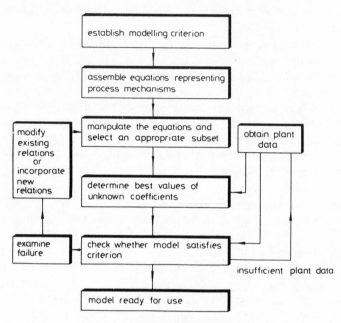

Fig. 1.2   *The modelling process*

1.3 Assembly of equations representing the basic plant mechanisms

The first stage in model building is to assemble equations representing all relevant mechanisms and components of the plant. Obviously, the equations will vary greatly according to the nature of the plant so that superficially there will appear to be little similarity between the equations representing different processes. However, in the initial search for basic equations the principles of *invariance* and *equilibrium* can be applied to the plant and to subsets of the plant. For a plant producing a product, the invariance principle leads to a set of *mass balance equations* of the general form

$$\text{Output mass } y = \text{input mass } u \pm \text{mass stored internally } q \qquad (1.1)$$

For a real plant, the set of equations representing this mass balance will be quite complex. Obviously, eqn. (1.1) can be differentiated to produce a relation between flow rates,

$$dy/dt = du/dt \pm dq/dt$$

Other results of application of the principle of invariance are:

A thermal balance equation

Heat generated internally
= sensible heat of products entering
− sensible heat of products leaving
± heat produced or absorbed by process $\quad$ (1.2)

An energy balance equation

Energy in system at time t
$\quad$ = energy in system at time 0
± energy dissipated or fed into the system
in the interval $[0, t]$ $\quad\quad$ (1.3)

A detailed balance of the energy in a system will often lead to a set of sound basic equations that can form the foundation for a model. This is to be expected since most dynamic systems can be characterized by their energy storage and energy dissipation behaviour.

The equilibrium equations include *Newton's law* for mechanical systems

$$\text{Applied force} = \text{mass} \times \text{acceleration} + \text{friction force} \quad (1.4)$$

$$\text{Applied torque} = \text{inertia} \times \text{angular acceleration}$$

$$+ \text{ friction torque} \quad (1.5)$$

while for electrical systems *Kirchhoff's laws* apply

The algebraic sum of currents at a node is zero $\quad\quad$ (1.6)

The sum of voltages around a closed circuit is zero $\quad\quad$ (1.7)

The above relations are of course very elementary but in a real situation there will be many such equations and the application of invariance and equilibrium principles leads to a structured approach to this initial stage of modelling. The initial equations will thus be the well known basic laws of physics and chemistry, particularly those related to invariance and equilibrium. For very simple plants, sufficient basic information may be generated by the above procedure to allow a model to be built. However, even for very simple plants there is usually at least one difficult relation that cannot easily be represented. Two examples illustrate this point. Consider the steering of a ship. It will be relatively easy to model the behaviour of the internal mechanisms of the ship, including the rudder dynamics. However, the effect on the ship's course of a change of rudder angle is obviously not so simple. At the very least it will depend on the speed and geometry of the ship. Consider next the simple case of a cold object suddenly inserted into a constant temperature oven held at 1000°C. Although the laws of heat radiation are directly applicable, the emissivity of the surface of the heated object is unlikely to be constant and it will have to be determined experimentally.

N.1.--B

	Generalized component	Electrical	Mechanical linear	Mechanical rotational	Thermal component	Hydraulic component	Pneumatic component
Resistive Component	$R = b/a$ Resistance	$R = v/i$ Resistance	$F = p/w$ Friction coefficient	$F = a/\omega$ Friction coefficient	$R = \theta/i$ Thermal resistance	$R = p/i$ Resistance	$R = p/i$ Resistance
Capacitive Component	$C = \frac{1}{a}\int b\, dt$ Capacitance	$C = \frac{1}{v}\int i\, dt$ Capacitance	$k = \frac{1}{p}\int w\, dt$ Spring constant	$k = \frac{1}{a}\int \omega\, dt$ Spring constant	$C = \frac{1}{\theta}\int i\, dt$ Thermal capacitance	$k = \frac{1}{p}\int i\, dt$ Compressibility	$k = \frac{1}{p}\int i\, dt$ Compressibility
Inertial Component	$J = \frac{a}{db/dt}$ Inertia	$L = \frac{v}{di/dt}$ Inductance	$M = \frac{p}{dw/dt}$ Mass	$J = \frac{a}{d\omega/dt}$ Inertia	Property does not exist	$J = \frac{p}{di/dt}$ Liquid inertia	$J = \frac{p}{di/dt}$ Gas inertia
Through variable	b	Current i	Velocity w	Angular velocity ω	Heat flow i	Liquid flow i	Gas flow i
Across variable	a	Voltage v	Force p	Torque a	Temperature difference θ	Pressure drop p	Pressure drop p

Fig. 1.3   Components and their analogues

The equations representing individual components of the plant will often be elementary and obvious. However, it is still useful to make use of the table of components and their analogues given in Fig. 1.3.

Although Fig. 1.3 and other elementary relations give a good foundation, a plant rarely satisfies the initial assumptions made at the start of the modelling procedure, and the modelling of process imperfections, non-linearities, non-homogeneity etc., requires detailed process knowledge and experimentation.

As an illustration, consider the rolling of hot metal strip between a single pair of parallel rolls.

Let H, h be the input and output thickness respectively. Let θ be the temperature of the strip in the roll gap. Let R be the radius of the rolls and S

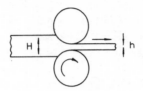

Fig. 1.4

the gap between the rolls when no strip is being rolled. Let ω be the angular velocity of the rolls, μ the coefficient of friction between the rolls and the strip and λ be a parameter representing the composition of material being rolled. Let M be the stiffness of the housing supporting the rolls and F be the separating force between the rolls.

The following equations can be found in the literature on metal rolling.

$$F = f(H, h, \theta, R, \omega, \mu, \lambda) \qquad \text{roll force equation} \quad (1.8)$$

$$h = S + F/M \qquad \text{Hooke's law} \quad (1.9)$$

These two equations can form the theoretical basis for a rolling model. However, even to satisfy relatively modest accuracy requirements, the following 'imperfections' must be taken into account.

1. The stiffness of the housing cannot be represented by a simple parameter M but rather by a curve that must be determined experimentally and stored.
2. The rolls are no longer parallel when subjected to forces generated by rolling.
3. The rolls are not perfectly circular due to wear and other imperfections.
4. The rolls are flattened when in contact with the strip so that the value of R for substitution into the roll force equation is not that of the unloaded rolls.
5. S is speed-dependent because of oil film effects in the roll bearings.

In addition, μ, θ and S are very difficult to measure for input to the model.

1.4 Constants, coefficients and variables

The model may contain one or more true *constants*, e.g. Planck's constant or the mechanical equivalent of heat. Such constants obviously present no problems.

Other coefficients may be *constant over the whole range of operation* of the model and for practical purposes they can also be regarded as constants. Examples are ambient temperature and physical dimensions of the plant.

Some coefficients may be considered *constant during any particular run of the model*. For instance in a batch process, certain coefficients will be constant for the duration of the batch but need resetting for each subsequent batch. In a continuous process, raw material density and composition can perhaps be considered constant during the lifetime of a particular consignment of raw material. When a new consignment of raw material comes into use, the appropriate coefficients must be modified. The discontinuity in the coefficient values causes a process disturbance that may have to be evaluated quantitatively by the model.

Finally, there are *coefficients that need to be modified during the running* of the model. Current values are determined either from a stored formula or by interpolation in a stored table. Examples are load torque as a function of speed and the inductance of an iron-cored coil as a function of current in the coil.

A dynamic model in the form of a vector-matrix differential equation

$$\dot{\mathbf{x}} = A\mathbf{x} + B\mathbf{u}$$

$$\mathbf{y} = C\mathbf{x}$$

has a natural classification of variables.

$\mathbf{x}$ is an n-dimensional vector of *state variables* where n is the order of the dystem.

$\mathbf{y}$ is an m-dimensional vector of *output variables*. In practice each of the elements of y must be determined by observation or measurement.

$\mathbf{u}$ is an r-dimensional vector of *input variables*. This vector must contain elements representing changing plant inputs, all known external disturbances that are to be represented, and the control variables.

An advantage of the vector-matrix formulation is that it leads naturally to the systematic classification of variables described above. Even if the model is not in a vector-matrix form, a similar type of classification of variables is still possible and desirable.

1.5 Modelling of distributed systems

Almost all physical effects are distributed in space rather than concentrated at a point. However, for many applications pointwise concentration can be assumed and this leads to a model containing ordinary differential equations. Systems having important spatially distributed effects must be represented by partial differential equations. Typical systems with distributed effects are furnaces (temperature distribution), reactors (composition distribution), transmission lines (current distribution) and rivers (composition distribution). A general form for the partial differential equation in one spatial dimension is

$$\frac{\partial x}{\partial t} + v \frac{\partial x}{\partial l} + r = 0 \tag{1.10}$$

x might be temperature within a moving column of liquid having velocity v in the l direction. r represents rate of heat generation in the column.

The *numerical solution of partial differential equations* requires discretization to produce a set of ordinary differential equations that are then solved by conventional methods for such sets of equations. Consider eqn. (1.10) above. $\partial x/\partial l$ can be written $(\partial x/\partial l)(t, l)$ since it depends on both time and distance along the column. However, for a sufficiently small spatial region about a point l_i so that $l_i - \epsilon \le l \le l_i + \epsilon$, $\partial x/\partial l$ can be considered to be a function of time only. Let

$$p_i(t) = \frac{\partial x}{\partial l}(t)$$

The spatial dimension can be split into n elements each of length 2ϵ so that the set of equations to be solved is

$$\frac{dx}{dt} + v p_i(t) + r = 0 \bigg|_{i = 1 \cdots n} \tag{1.11}$$

Examination of what has just been done shows that it is equivalent to the physically based assumption that the moving column can be partitioned into n zones—within each zone the variation of x with l can be assumed constant at any particular time t.

Thus the discretization of distributed systems, which is a necessary step in their modelling, can be undertaken on physical grounds by partitioning into spatial zones in the model formulation stage. Alternatively, discretization can be undertaken on a purely mathematical basis at a later stage. The two approaches lead to a similar set of ordinary differential equations to be solved. However, for a complex set of partial differential equations it may not be possible, where the second approach is used, to give a physical meaning to the discretization. As a second illustration of discretization, consider the equation which arises in two-dimensional temperature diffusion.

$$\frac{\partial^2 \theta}{\partial y, \partial z} = \frac{1}{k} \frac{\partial \theta}{\partial t} \tag{1.12}$$

where θ = temperature and k is a constant, and y, z are the two spatial variables.

The variable $\partial^2\theta/\partial y,\ \partial z$ can be spatially discretized as follows.

$$\frac{\partial^2\theta}{\partial y,\ \partial z} = \frac{\partial}{\partial y}\left(\frac{\partial\theta}{\partial z}\right) \simeq \frac{\partial}{\partial y}\left(\frac{\theta(y_0, z_0 + \epsilon) - \theta(y_0, z_0 - \epsilon)}{2\epsilon}\right)$$

$$\simeq \left(\frac{\theta(y_0 + \epsilon, z_0 + \epsilon)}{2\epsilon} - \frac{\theta(y_0 - \epsilon, z_0 + \epsilon)}{2\epsilon}\right)\frac{1}{2\epsilon}$$

$$- \left(\frac{\theta(y_0 + \epsilon, z_0 - \epsilon)}{2\epsilon} - \frac{\theta(y_0 - \epsilon, z_0 - \epsilon)}{2\epsilon}\right)\frac{1}{2\epsilon}$$

Thus

$$\left.\frac{\partial^2\theta}{\partial y,\ \partial z}\right|_{y_0, z_0} \simeq \frac{\theta(y_0 + \epsilon, z_0 + \epsilon) - \theta(y_0 - \epsilon, z_0 + \epsilon)}{4\epsilon^2}$$

$$- \frac{\theta(y_0 + \epsilon, z_0 - \epsilon) - \theta(y_0 - \epsilon, z_0 - \epsilon)}{4\epsilon^2}$$

This approach to the modelling of two-dimensional temperature distribution requires a spatial grid to be drawn. At each node of the grid a different ordinary differential equation needs to be solved. See Fig. 1.5.

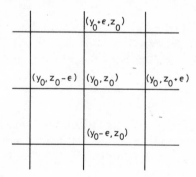

Fig 1.5

The modelling of distributed systems is in practice very difficult. Discretization can produce misleading spurious effects in the model, owing to the artificial discontinuities that are introduced at the boundary of spatial zones. In some applications, such as the modelling of melting and solidification, the spatial boundary is changing with time so that the discretization grid may have to be expanded as the solution proceeds.

1.6 Modelling of process dead-time

Consider the simple arrangement of Fig. 1.6

The block takes in a function $f(t)$ and gives out the same function delayed in time by τ time units, but otherwise, unaltered.

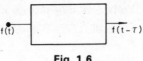

Fig. 1.6

When the block diagram of a process contains a block like that of Fig. 1.6 the process is said to have *dead-time, transport lag* or *finite delay*—the terms being equivalent.

From the point of view of control, dead-time is highly undesirable. Two useful references are the early paper[3] by Callender and the book.[17] Process configurations that have dead-time are the following:

(1) In general, processes where a product flows from one place to another.
(2) Processes where measuring sensors are situated some distance 'downstream' of the process proper, thus causing a dead-time in the measurement. From the point of view of control this dead-time can be considered part of the process itself.
(3) (Similar to 2) Processes whose output is monitored by some special sensor, such as a chemical analyser, that gives out its result delayed by τ time units.

It may not be easy to distinguish the presence of dead-time, simply by looking at the step reponse of a process.

Fig. 1.7 shows sketches to help the visualization of the above point.

Fig. 1.7 *Comparison of the step responses of processes (i) and (ii).*
(i) A low-order dynamic process G_1 in series with a dead-time.
(ii) A high-order dynamic process G_2 with no dead-time.

Usually, by physical reasoning, it will be obvious which of the cases applies. However, we can make use of this similarity and, if it suits our purpose, represent a finite delay by a higher-order process. [In analogue computation where finite delays are difficult to produce, the so-called Padé approximation does exactly this. Conversely, we can represent a high-order system by a dead-time in series with a low-order system. Such an approach was used in the early work by Ziegler and Nicholls,[25] in the simple modelling of processes to allow the setting-up of commercial controllers.] For instance, in the modelling of an industrial oven, it will often be sufficient, for purposes of subsequent control, to represent the process by a low-order dynamic system in series with a dead-time. If the model is to be used for oven design rather than control, we shall have to take a more fundamental approach involving the partial differential equations of heat diffusion, ensuring that parameters of physical relevance appear in the model.

Representation of dead-time in a model

Considered as a transfer function, the dead-time block of Fig. 1.6 is represented by the transfer function $G(s) = e^{-\tau s}$. Since $e^{-\tau s}$ can be expanded in a series which can then be truncated, it is easy to see how dead-time can be approximated by a sufficient number of terms in s.

In the representation of a process by differential equations, the implicit assumption is that future behaviour is uniquely determined by the present state and future inputs, and is independent of the past. In the presence of dead-time, the assumption clearly fails and ordinary differential equations do not suffice to describe the process. To take the finite delay into account, we are led to differential-difference equations. Typical differential-difference equations are

$$\dot{x}(t) = x(t - \tau) \qquad (1.13)$$

$$\dot{x}(t) = f(x(t), u(t - \tau)) \qquad (1.14)$$

The first equation is a linear-delay equation while the second is a more general non-linear example. Such equations appear frequently in mathematical economics and mathematical biology as well as in engineering.

Even eqn. (1.13), which appears to be simple, poses problems in that for its solution we need to specify not an initial condition but an initial function $x(t)$ over the interval $[-\tau, 0]$. Differential-difference equations belong to the wider class of functional-difference equations. An excellent text on differential equations that includes a treatment of delay equations is Reference 6. A more abstract book that covers functional differential equations is Reference 9.

Numerical modelling of dead-time is trivially achieved in digital simulation by simply storing sampled variables in a moving array of the right dimension. The only problems are concerned with logistics and quantity of storage required.

Variable dead-time

It is unusual to find a process dead-time that is really constant. For instance, in a process where the dead-time is due to the time taken for a moving product to reach a measuring sensor, any variation in product velocity will alter the dead-time. In general, for a linear process, we shall have the situation represented by the equations

$$\dot{\mathbf{x}}(t) = A\mathbf{x}(t) + B\mathbf{u}(t)$$

$$\mathbf{y}(t) = C\mathbf{x}(t - \tau(\mathbf{x})) \qquad (1.15)$$

the notation indicating that the time delay is a function of the state vector of the process.

Practical considerations

Although the process dead-time can be estimated from plant data just as for any other process parameter, it will often be advantageous to use cross-correlation for this purpose.

Cross-correlation can be carried out off-line on recorded data but there exist commercial correlators that are ideally suited for use on the process, while it is in normal operation, to determine dead-time and other dynamic characteristics.

1.7 Experimental approaches to model building

Where a model of an existing plant is to be constructed, a set of input-output data can be obtained and mathematical relations can be established, without regard to the physical structure of the plant, that reproduce the same input-output behaviour as that of the plant. In the simple case of a linear noise-free low-order single-input single-output system, a step-response or frequency-response curve can be recorded and a transfer function or differential equation chosen that gives a good approximation to the recorded response.

Interactive software packages exist to allow rapid modelling of the type described above. It is always difficult in the absence of prior knowledge to decide what order of model to take, given only input-output data. The interactive facility allows the user to choose between models of different order. An alternative approach to modelling, which again ignores the physical structure of the plant, is to determine the mathematical relations that statistically give the best fit to the recorded data. These methods of statistical modelling are briefly reviewed in Section 1.10.

1.8 Experimental determination of unknown coefficients in models—parameter estimation

The principle of parameter estimation is illustrated below. Plant input-output data are recorded. The model, containing parameters to be determined, is subjected to the recorded input data and the parameters varied until the model output is as nearly as possible the same as that recorded at the plant. An automatic method for determining the best parameters is known as a *hill-climbing* procedure.

The data, recorded over a time interval $[0, T]$ are repeatedly input to the model to give different values of J. When the hill-climbing procedure cannot reduce the scalar-valued criterion J further, the vector $\hat{A}$ is taken to be the best estimate of the plant parameters. The principle of hill climbing is shown in Fig. 1.8.

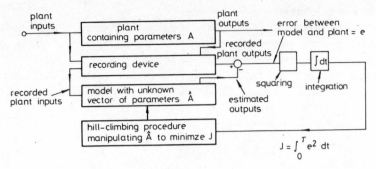

Fig. 1.8

Assume that there are two unknown parameters a_1, a_2 so that J is a function of a_1, a_2 ($J = J(a_1, a_2)$). A simple hill-climbing procedure as shown in the flow diagram of Fig. 1.11 will quickly determine the minimum value of J provided that the J contours are roughly circular (Fig. 1.9). When, as is

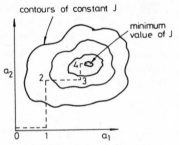

Fig. 1.9 *Hill climbing on a circular hill*

commonly the case, the contours are far from circular, the simple hill-climbing procedure no longer succeeds, the process terminating on a ridge well away from the minimum value of J as illustrated in Fig. 1.10. All practical hill-climbing methods use either ridge-following techniques or so modify the scaling of the unknown parameters that the J contours become more nearly circular.

The simple program shown in Fig. 1.11 manipulates one coefficient at a

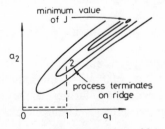

Fig. 1.10 *Hill climbing on a ridge*

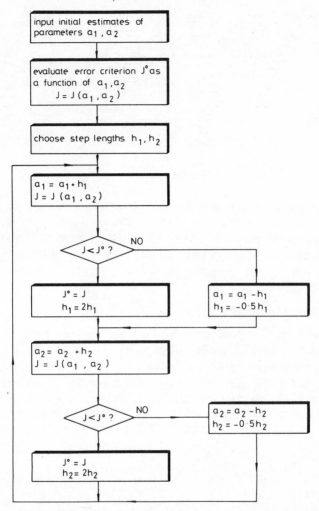

Fig. 1.11 *Simple hill-climbing program to determine the best values of two coefficients, a_1, a_2*

time and works well only on problems with nearly circular J contours. At any time, $J°$ is the lowest value so far attained for the error criterion.

Figure 1.12 illustrates how the error criterion J might be determined for a plant with two outputs y_1 and y_2. In the case illustrated J would have the form

$$J = \int_0^T \{\gamma_1(y_1 - \hat{y}_1)^2 + \gamma_2(y_2 - \hat{y}_2)^2\} \, dt$$

with the weighting factors γ_1, γ_2 being chosen to take into account scaling of the variables y_1, y_2 and the relative importance to the model of these variables.

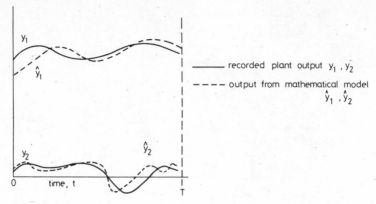

Fig. 1.12 *Illustrating how the error criterion* $J = \int_0^T e^2 \, dt$ *in fig. 1.11 is typically determined*

There are many specialist parameter estimation techniques and Fig. 1.13 gives an indication of the available methods and their application. Fig. 1.13 is based on a much larger diagram in Reference 8.

1.9 Linearization techniques

During modelling, non-linear relations are often replaced by linear approximation in the interests of mathematical tractability of the model. Obviously, nearly linear functions might be replaced by straight-line approximations. *Piecewise linearization* splits the function into a number of linear approximations. As the model operates and moves into different regions, the approximations used have to be changed. Where the non-linear relations are defined by experimental data, linearization can be carried out by regression. Where the non-linearity is defined analytically, it can be *expanded in a Taylor series* in which all but the linear terms are neglected.

Let $x = f(u)$ be a differentiable non-linear relation between u and x.

Expanding about any chosen point u_0 in a Taylor series gives

$$x = f(u_0) + (u - u_0)\frac{df}{du}\bigg|_{u_0} + \text{higher order terms} \qquad (1.16)$$

Let $x_0 = f(u_0)$ then

$$x_0 + x - x_0 = f(u_0) + (u - u_0)\frac{df}{du}\bigg|_{u_0} + \text{higher order terms} \qquad (1.17)$$

$$x - x_0 = (u - u_0)\frac{df}{du}\bigg|_{u_0} + \cdots$$

or $$\delta x = \delta u \frac{df}{du}\bigg|_{u_0} + \text{higher order terms}$$

where $(\delta u, \delta x)$ is the perturbation from the chosen linearization point (u_0, x_0).

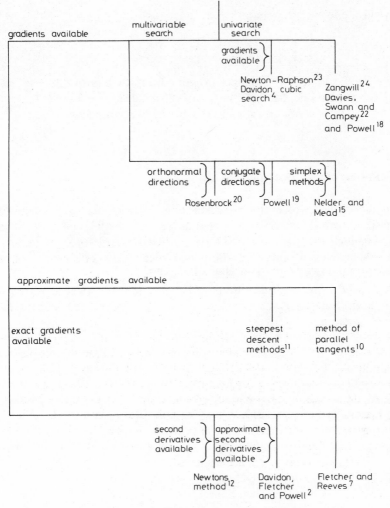

Fig. 1.13 *An introduction to hill-climbing methods*

If f is a function of several variables, a Taylor's series can again be used. Thus let

$$x = f(u_1, u_2, \ldots, u_n) \qquad u = (u_1, u_2, \ldots, u_n)$$

then Taylor's series expansion gives

$$x = f(u_1^\circ, \ldots, u_n^\circ) + (u_1 - u_1^\circ) \frac{\partial f}{\partial u_1}\bigg|_u$$

$$+ \cdots + (u_n - u_n^\circ) \frac{\partial f}{\partial u_n}\bigg|_u + \text{higher order terms} \qquad (1.18)$$

To yield the linear approximation

$$\delta x = \delta u_1 \left. \frac{\partial f}{\partial u_1} \right|_u + \cdots + \delta u_n \left. \frac{\partial f}{\partial u_n} \right|_u \tag{1.19}$$

where $(\delta u_1, \cdots, \delta u_n, \delta x)$ is the perturbation from the point $(u_1^{\circ}, \cdots, u_n^{\circ}, x^{\circ})$.

In performing linearization it is always advisable to check the magnitude of the higher-order terms to estimate the degree of approximation that is being made.

1.10 Review of statistical methods

Since modelling involves the interpretation of experimental data, a knowledge of basic statistical methods is a prerequisite for successful modelling. In particular, *planning of experiments* and *tests of significance* are topics of importance that are pursued in many standard texts. Below is a brief review of other statistical methods of use in modelling.

Scaling of variables. It is good practice to scale variables so that they have zero mean and unity variance. In statistics this is referred to as *zero-one normalization*.

Regression. Given a set of measurements $\{x, u_1, \ldots, u_n\}$, where it is assumed that $x = f(u_1, \ldots, u_n)$, regression is concerned with determining the function $\hat{f}$ such that $J = \sum_{i=1}^{n} (x_i - \hat{x}_i)^2$ is minimized where $\hat{x} = \hat{f}(u_1, \ldots, u_n)$ and where there are n samples of measured data available. Assume that the function $\hat{f}$ is determined by m unknown parameters then a measure of the fit of the function $\hat{f}$ to the function f is given by the *standard error*

$$E = \left[\frac{J}{n - m} \right]^{1/2} \tag{1.20}$$

In *linear regression* the function f is restricted to the form

$$f(u_1, \ldots, u_n) = a_0 + a_1 u_1 + \cdots + a_n u_n$$

and the expression for J can be differentiated successively with respect to the a_i to yield equations that can be solved analytically.

Residual analysis. Residual analysis is concerned with the examination of plots of $(x - \hat{x})$, i.e. of the error between model output and recorded plant output, against time or against some other variable. Any deterministic trend in such plots indicates the presence of an effect that has not been adequately modelled. Davidson,[5] describes this procedure in detail.

Covariance and correlation. Let $\{x_i\}\{y_i\}$ be two zero-one normalized time series, thus, the *covariance* between x and y is defined by

$$\text{cov}(x, y) = \frac{1}{n} \sum_{i=0}^{n} x_i y_i \tag{1.21}$$

The correlation coefficient between x and y is defined by

$$R(x, y) = \text{cov } (x, y)/(\sigma^2(x)\sigma^2(y)) \tag{1.22}$$

where $\sigma^2(x)$, $\sigma^2(y)$ indicate the variances of the time series $\{x_i\}$, $\{y_i\}$.

Model reduction. Suppose that a model contains two variables x and y and that the correlation coefficient between x and y is unity, then x and y are virtually identical within a constant of proportionality, and one can be eliminated and the complexity of the model consequently reduced. In case x and y are highly correlated, *principal component analysis* can be used to define two new orthogonal variables.

Let $\{x\}$, $\{y\}$, $\{z\}$ be three time series to which principal component analysis is to be applied.

Let C be the matrix defined by

$$C = \begin{pmatrix} R(x, x) & R(x, y) & R(x, z) \\ R(y, x) & R(y, y) & R(y, z) \\ R(z, x) & R(z, y) & R(z, z) \end{pmatrix} \tag{1.23}$$

where the R are correlation coefficients.

Let e_i; $i = 1, 2, 3$ be the eigenvectors of the matrix C.

Let λ_i; $i = 1, 2, 3$ be the eigenvalues of the matrix C.

The new transformed variables generated by the principal component analysis are the eigenvectors e_i. Any e_i for which the corresponding λ_i is very small can be neglected and this leads to model reduction. The procedure of principal component analysis is described in detail in Reference 11.

Coefficient invariance. If a model contains redundant variables or highly correlated variables then large changes can be made in some coefficient values without significantly altering the behaviour of the model, i.e. a wide range of coefficient values give the same performance from the model. In this situation small changes in the plant give rise to large changes in the coefficients of the model. A model of a time invariant plant should not change its coefficients greatly, when fitted against different sets of plant data. Any significant change should be investigated and redundant variables and strong correlations between variables eliminated.

1.11 Solution of ordinary differential equations in the digital computer

For multivariable linear models whose dynamics can be represented by the vector matrix equation $\dot{x} = Ax + Bu$ the time solution is

$$\mathbf{x}(t) = e^{At}\mathbf{x}(0) + \int_0^t e^{A(t-\tau)}\mathbf{Bu}(\tau) \, d\tau$$

For simulation purposes the equation can be written in discrete time form

$$\mathbf{x}_{(n+1)T} = e^{AT}\mathbf{x}_{nT} + \int_{nT}^{(n+1)T} e^{A((n+1)T-nT-\tau)}\mathbf{Bu}(\tau) \, d\tau$$

$$= e^{AT}\mathbf{x}_{nT} + \int_{nT}^{(n+1)T} e^{A(T-\tau)}\mathbf{Bu}(\tau) \, d\tau \tag{1.24}$$

T is a time step, and if T is sufficiently small, the variable u can be considered constant yielding

$$\mathbf{x}_{(n+1)T} = e^{AT}\mathbf{x}_{nT} + [-A^{-1}e^{AT}e^{-A\tau}\mathbf{Bu}]_{nT}^{(n+1)T}$$

$$\mathbf{x}_{(n+1)T} = e^{AT}\mathbf{x}_{nT} + A^{-1}[e^{AT}-I]\mathbf{Bu}, \qquad \text{provided det } A \neq 0 \tag{1.25}$$

$\mathbf{u}$ takes a single value, assumed constant, valid for the interval $[nT, (n+1)T]$, I is the identity matrix. Eq. (1.25) can be solved repeatedly in time steps of length T to yield the time solution of the original vector-matrix differential equation.

Determination of e^{AT} for solution of eqn. (1.25)
e^{AT} can be determined by the series expansion

$$e^{AT} = I + AT + \frac{A^2T^2}{2!} + \cdots \tag{1.26}$$

which is always convergent and quickly convergent provided that the time step T is sufficiently short.

e^{AT} is known as the transition matrix and there are several methods for its calculation that are more attractive than the above series expansion. A Fortran program that calculates e^{AT} by Sylvester's method can be found in Melsa.[14] A good text covering the solution of the equation $\dot{\mathbf{x}} = A\mathbf{x} + B\mathbf{u}$ is Ogata.[16]

For systems of nonlinear differential equations, the basic numerical method of solution is the *Runge-Kutta method*. There are several variants of the method but the principle is as follows. Let the non-linear differential equation have the form

$$\frac{dx}{du} = f(x, u) \tag{1.27}$$

If x is a function of u then $x(u)$ can be expanded in the Taylor series

$$x(u) = x_0 + \frac{dx}{du}(u - u_0) + \frac{1}{2}\frac{d^2x}{du^2}(u - u_0)^2 + \cdots \tag{1.28}$$

Define $h = u - u_0$, then

$$x(u_0 + h) = x_0 + h\frac{dx}{du} + \frac{h^2}{2}\frac{d^2x}{du^2} + \cdots \tag{1.29}$$

Given an initial condition (x_0, u_0), $x(u_0 + h)$ can be calculated and the process can be repeated so that the values of x at discrete intervals of length h in u can be determined.

The Runge-Kutta method uses the following relations to approximate the first four terms of the Taylor series expansion. Let

$$k_1 = hf(x_0, u_0)$$

$$k_2 = hf\left(x_0 + \frac{k_1}{2}, u_0 + \frac{h}{2}\right)$$

$$k_3 = hf\left(x_0 + \frac{k_2}{2}, u_0 + \frac{h}{2}\right)$$

$$k_4 = hf(x_0 + k_3, u_0 + h) \tag{1.30}$$

Then

$$x(u_0 + h) = x_0 + \tfrac{1}{6}(k_1 + 2k_2 + 2k_3 + k_4) \tag{1.31}$$

This is a fourth order Runge-Kutta method since eqn. (1.31) represents the first four terms of a Taylor series expansion.

There are several other numerical methods rivalling the Runge-Kutta method. Notably, the so-called predictor-corrector methods are generally faster than Runge-Kutta methods although they are not self-starting, and this introduces complexity.

1.12 Reasons for model failure

Even after considerable expenditure of effort, it happens that a model may not meet the specification laid down for it. Failure can occur due to:

(i) Lack of sufficient relevant plant data with which to develop the model.

(ii) Failure to obtain a well-structured foundation to support the later stages of model refinement—this is due to lack of expertise or lack of process knowledge.

(iii) Numerical or computational problems of model implementation including

— failure of convergence of iterative loops

— excessive time requirements of computation so that the model is virtually useless

Failure of convergence of iterative loops

The principle is as follows. Two equations that are too complex to be combined are solved separately to yield an iterative process. Let the two equations be

(a) $y = mx + c$

(b) $y = lx + d$ or $x = l^{-1}(y - d)$

An initial estimate x_0 is inserted into eqn. (a) leading to an estimate y_0.

y_0 inserted into eqn. (b) produces an estimate x_1, which, inserted into eqn. (a) produces an estimate y_1. This procedure is continued until x_n, y_n both satisfy the equation as closely as desired. The procedure can be illustrated graphically, as in Fig. 1.14.

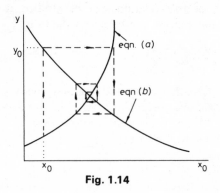

Fig. 1.14

The convergence of this and other iterative processes is best studied not by graphical methods but by using the fixed-point theorem as described below. Let an iterative process be represented in general by the relation

$$x_{n+1} = Ax_n \tag{1.32}$$

where the operator A represents the iteration process and where x may be a vector with components $x_1, x_2, ..., x_n$.

A is defined to be a *contraction operator* if

$$\|A\mathbf{x}\| \le k\|\mathbf{x}\|$$

where k satisfies $k \le 1$ and $\|\mathbf{x}\|$ is defined by $\|\mathbf{x}\| = (x_1^2 + x_2^2 + \cdots x_n^2)^{1/2}$. In case x is a scalar variable, A is defined to be a contraction operator if

$$|Ax| \le k|x|$$

where k satisfies $k < 1$ and $|x|$ denotes the modulus or absolute value of x.

Fixed-point theorem
Let A be a contraction mapping and further let A satisfy the condition $|Ax_1 - Ax_2| \le M|x_1 - x_2|$ for some positive constant M and for any x_1, x_2. (This is known as a Lipschitz condition.)

Then the recursive equation $x_{n+1} = Ax_n$ is convergent to a unique element x.

Example
Let the following equations be solved recursively using Newton's method (described in Reference 12).

$y = a \ln x$ ⎱ Let $F = a \ln x - kx$ then $F = 0$
$y = kx$ ⎰ implies that x satisfies both equations

Newton's method leads to the recursive equation

$$x_{n+1} = x_n - F_n/F'_n$$

where $F'_n = (d/dx)(F_n)$.

The convergence of this relation can be studied using the fixed-point theorem. The iteration process can be put in the form $x_{n+1} = Ax_n$ where

$$Ax_n = x_n - \left(\frac{a \ln x_n - kx_n}{a/x_n - k} \right) \qquad (1.33)$$

The operator A is continuous and satisfies the Lipschitz requirement, hence, if A is a contraction operator the iteration procedure will converge to a unique x. By substituting in the values for a, k, and considering the range within which the value of x must lie, information on the convergence properties can be obtained.

1.13 Conclusions

The stages in developing a mathematical model are:

1. To define the scope and accuracy of the desired model.
2. To write down a framework of theoretically based equations that, as far as possible, cover all the aspects of first-order importance.
3. If necessary, to supplement the framework obtained in (2) by equations based on experiment and observation.
4. To manipulate the framework of equations into a form suitable for solution in the computer.
5. To plan and carry out a data acquisition exercise.
6. To determine unknown parameters in the model.
7. To obtain further data allowing validation of the model.

For all but very simple modelling problems, a great deal of iteration will be required in the above scheme before the aims set down in (1) can be met.

The modeller needs to be skilful in formulating meaningful equations, in determining unknown parameters and in solving the equations efficiently. The formulation of meaningful equations can rarely be reduced to a systematic procedure but a great deal can be learnt from case studies such as the ones that follow in later chapters.

Suggestions for further reading
Lee[13] contains practically orientated material relevant to process control modelling.

Beveridge[1] is a very good reference for parameter estimation methods. Hyvarinen[11] is recommended for statistical aspects.

1.14 References

1 BEVERIDGE, G. D. C., and SCHECHTER, R. J.: 'Optimisation—theory and practice' (McGraw-Hill, 1970)
2 BROYDEN, C. G.: 'Quasi-Newton methods and their application to function minimisation', *Math. Comput.*, 1967, **21**, pp. 368–381
3 CALLENDER, A., *et al.*: 'Time lag in a control system', *Phil. Trans. R. Soc. London*, 1936, **A235**, pp. 415–444
4 DAVIDON, W. C.: 'Variable metric method for minimisation. Res. and Dev. Rep., 1969, *ANL-5990, A.E.C.*
5 DAVIDSON, H.: 'Statistical methods in model development', *in*, 'Computer Control of Industrial Processes', SAVAS, E. S. (Ed.) (McGraw-Hill, 1965)
6 DRIVER, R. D.: 'Ordinary and delay differential equations', (Springer, 1977)
7. FLETCHER, R., and REEVES, C. M.: 'Function minimisation by conjugate gradients', *Comput. J. (GB)*, 1964, **7**, pp. 149–154
8 GHANI, S. N., and BARNES, L.: 'Parameter optimisation for unconstrained object functions—a bibliography', *Comput. Aided Des.*, Oct. 1972, **4**(5), pp. 247–260
9 HALE, J.: 'Functional differential equations' (Springer, 1971)
10 HARKIN, A.: 'The use of parallel tangents in optimisation. Optimisation Techniques', pp. 35–40, (Blakemore and Davies, 1964)
11 HYVARINEN, L.: 'Mathematical modelling for industrial processes' (Springer, 1970)
12 KREYSZIG, E.: 'Advanced engineering mathematics' (John Wiley, 1972)
13 LEE, T. H., ADAMS, G. E., and GAINES, W. M.: 'Computer process control—modelling and optimisation' (John Wiley, 1968)
14 MELSA, J. L.: 'Computer programs for computational assistance in the study of linear control theory' (McGraw-Hill, 1975)
15 NELDER, J. A., and MEAD, R.: 'A simple method for function minimisation', *Comput. J. (GB)*, 1965, **7**, pp. 308–313
16 OGATA, K.: 'State space analysis of control systems' (Prentice-Hall, 1967)
17 OGUZTÖRELI, M. N.: 'Time lag control systems' (Academic Press, 1966)
18 POWELL, M. J. D.: 'An efficient method of finding the minimum of a function of several variables without calculating derivatives', *Comput. J. (GB)*, 1964, **7**, pp. 155–162
19 POWELL, M. J. D.: 'An efficient method of finding the minimum of a function of several variables without calculating derivatives', *Comput. J. (GB)*, 1964, **7**, pp. 303–307
20 ROSENBROCK, H. H.: 'An automatic method for finding the greatest and the least value of a function', *Comput. J. (GB)*, 1960, **3**, pp. 175–184
21 ROSENBROCK, H., and STOREY, C.: 'Computational techniques for chemical engineers' (Pergamon Press, 1966)
22 SWANN, W. H.: 'Report on the development of a new direct search method of optimisation', Central Instrument Laboratory Research Note 64/3, 1964, ICI Ltd.
23 WEISS, E. A., ARCHER, D. H., and BURT, D. A.: *Petr. Refiner*, 1961, **40**(10), pp. 169–174
24 ZANGWILL, W. I.: 'Minimising a function without calculating derivatives', *Comput. J. (GB)*, 1967, **10**, pp. 293–296
25 ZIEGLER, J. G., and NICHOLLS, N. B.: 'Optimum settings for automatic controllers', *ASME Trans.* 1942, **64**(8)

Chapter 2

Modelling of chemical process plant

J. B. Edwards

2.1 Introduction

A chemical plant involves the following basic operations, or stages:

(a) raw-material handling and preparation
(b) the chemical reaction itself and
(c) the separation of the various saleable products from one another and from the waste products.

In practice, however, the three operations may not take place separately. The rate of a chemical reaction, for instance, may be determined not so much by chemical kinetics as by the mechanical processes of mixing the reagents (the reacting components) within the reactor if these are not freely miscible. Fluid and solid mechanics may therefore dominate even the chemical reaction stage (b) of production in some instances. In other situations, thermodynamics may play the dominant role in the reactor since the velocity of chemical reaction is often highly temperature-dependent and furthermore, large quantities of heat can be generated or absorbed in the course of the reaction. Thermal considerations are obviously of paramount importance in such cases. Conversely, the separation stage (c), although frequently involving the thermodynamic process of distillation or perhaps the fluid mechanical process of solvent extraction, may nevertheless be affected by the continuing reaction of unused reagents within the separation vessel after these have left the reactor proper.

As a final example of the difficulty of categorizing real-life operations we should also note that even the preparation stage (a) may also involve more than just the grading and premixing of reagents: as in the case of preparing sinter feeds for the blast furnace where mechanical transportation and combustion take place simultaneously.

The isolation of three sequential stages in chemical production is therefore only an idealizing concept. Likewise the basic physical and chemical phenomena (or processes) involved, viz.

(i) mass transfer
(ii) heat transfer and
(iii) chemical change

do not take place completely separately and cannot in general be associated uniquely with any of the three production stages (a), (b) or (c). It is nevertheless essential to study idealized systems dominated by single isolated processes to acquire insight into their behaviour before proceeding to large-scale simulation or pilot-plant studies of systems involving all three phenomena operating in parallel because of the risk of error (from human, numerical or instrumentation sources) and the often unmanageable number of degrees of freedom otherwise presented by full-scale process simulation.

This chapter therefore sets out to show how idealized unit processes might be modelled analytically so that expected approximate solutions may be generated against which full-scale system simulations can be tested.

2.2 Setting up the process equations

Process equations are formulated from fairly elementary concepts of physical and chemical dynamic balance and equilibrium. Three system examples are considered here, each dominated by a different process from the list (i), (ii), (iii) (above). We consider first an elementary, liquid/liquid heat exchanger of the counterflow type.

2.2.1 Heat-exchanger example

The system is illustrated diagrammatically in Fig. 2.1. The two liquids flow in opposite directions parallel to one another at mass rates W_1 and W_2 and separated by a heat-conducting interface which will be assumed to have negligible thermal capacitance or resistance. The outer shell is assumed to be perfectly insulating. For modelling purposes the process is first imagined to be subdivided into N cells each of length $\delta h'$ and all identical except that cells 1 and N have atypical conditions pertaining at their left- and right-hand boundaries, respectively. (These are the process boundary conditions to be discussed later.) Within each cell, conditions are assumed to be completely homogeneous either side the interface so that the temperatures of the two fluids 1 and 2 within cell n may be represented by the single variables $\theta_1(n, t)$ and $\theta_2(n, t)$, respectively. The functions $\theta_1(h', t)$ and $\theta_2(h', t)$, where h' denotes a general distance from the left-hand process boundary, are therefore approximated initially by spatially discrete functions $\theta_1(n, t)$ and $\theta_2(n, t)$, $n = 1, 2, \ldots, N$, which undergo step change at each cell boundary. This is an example of the use of the 'stirred-tank concept' employed in process modelling, each

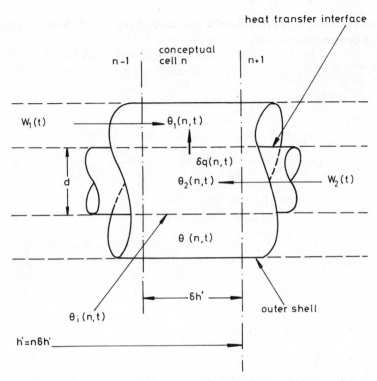

Fig. 2.1   *Variables associated with nth cell of liquid/liquid heat exchanger*

cell being thought of as a discrete tank, the contents of which are thoroughly mixed.

There being no chemical change in this process, it is now necessary merely to draw up an inventory of the material and heat entering and leaving each cell (or tank) compartment, any imbalance between inflow and outflow being equated to a rate of buildup of material or heat within the compartment. The material balance is trivial in this example (if the fluids are assumed to be incompressible, of constant density and completely filling the cells at all times), yielding merely the result that W_1 and W_2 are functions of time only and invariant in h'. The heat balance for the system is more complicated however and, for fluids 1 and 2 in cell n, may be written

$$S_1 \rho_1 A_1 \, \delta h' \, d\theta_1(n, t)/dt = W_1 S_1 \{\theta_1(n-1, t) - \theta_1(n, t)\} + \delta q(n, t)$$
$$S_2 \rho_2 A_2 \, \delta h' \, d\theta_2(n, t)/dt = W_2 S_2 \{\theta_2(n+1, t) - \theta_2(n, t)\} - \delta q(n, t) \qquad (2.1)$$

where the suffixes apply to the associated fluids and compartments, S denotes specific heat, ρ fluid density and A the cross-sectional area of the flow passage. $\delta q(n, t)$ is the rate of heat flow across the interface from liquid 2 to liquid 1 and is proportionally dependent upon the temperature difference between the liquids and the interface wall and also upon the interface area $\pi d \; \delta h'$ (where d is the interface tube diameter). The major resistance to heat flow is provided by thin boundary layers of near-stationary liquid, lining both sides of the interface, the thicknesses of which are found to decrease with flow rate. In fact it is found that this resistance is proportional to $W^{-0.8}$ so that

$$\delta q(n, t) = k_1 \pi d W_1^{0.8}\{\theta_i(n, t) - \theta_1(n, t)\} \; \delta h'$$

$$= k_2 \pi d W_2^{0.8}\{\theta_2(n, t) - \theta_i(n, t)\} \; \delta h'$$

where k_1 and k_2 are constant coefficients of heat transfer. θ_i denotes interface temperature, and may be eliminated from the two equations above to give

$$\delta q(n, t) = \frac{k_1 k_2 \pi d (W_1 W_2)^{0.8}\{\theta_2(n, t) - \theta_1(n, t)\} \; \delta h'}{k_1 W_1^{0.8} + k_2 W_2^{0.8}} \qquad (2.2)$$

Given the constant parameters of the system and input functions $W_1(t)$, $W_2(t)$, $\theta_1(0, t)$ and $\theta_2(N, t)$ eqns. 2.1 and 2.2 are suitably expressed for computer simulation of the system provided a value for N can be specified for solutions of the desired accuracy. For analytical solution however we require a much more compact model representation.

Now the true temperature functions will be continuous in time and space between the boundaries (at $h' = 0$ and L' in this case) and will therefore be governed by Taylor series so that

$$\theta_1(n, t) - \theta_1(n - 1, t) = \frac{\partial \theta_1(n, t)}{\partial h'} \; \delta h'$$

$$- \frac{1}{2} \frac{\partial^2 \theta_1(n, t)}{(\partial h')^2} (\delta h')^2 + \text{higher powers of } \delta h'$$

$$\theta_2(n + 1, t) - \theta_2(n, t) = \frac{\partial \theta_2(n, t)}{\partial h'} \; \delta h'$$

$$+ \frac{1}{2} \frac{\partial^2 \theta_2(n, t)}{(\partial h')^2} (\delta h')^2 + \text{higher powers of } \delta h'$$

$$(2.3)$$

and as $\delta h'$ is made progressively smaller for a given length L' of process, i.e. as N, $(= L'/\delta h')$ is made progressively larger, the approximate discrete temperature functions governed by eqn. (2.1) and the continuous functions governed by eqn. (2.3) will tend to equality so that we are justified in substituting eqn. (2.3) in (2.1) if $\delta h' \to 0$, yielding:

$$S_1 \rho_1 A_1 \frac{\partial \theta_1}{\partial t} = - W_1 S_1 \frac{\partial \theta_1}{\partial h'} + \frac{k_1 k_2 \pi d (W_1 W_2)^{0.8}}{k_1 W_1^{0.8} + k_2 W_2^{0.8}} (\theta_2 - \theta_1)$$

$$S_2 \rho_2 A_2 \frac{\partial \theta_2}{\partial t} = W_2 S_2 \frac{\partial \theta_2}{\partial h'} - \frac{k_1 k_2 \pi d (W_1 W_2)^{0.8}}{k_1 W_1^{0.8} + k_2 W_2^{0.8}} (\theta_2 - \theta_1),$$

$$(2.4)$$

$$0 < h' < L'$$

Later in the chapter we shall consider further a symmetrically built and operated process, i.e. one in which $A_1 = A_2 = A$, $k_1 = k_2 = k$, $S_1 = S_2 = S$, $\rho_1 = \rho_2 = \rho$, operated under the *nominal* working condition

$$W_1 = W_2 = W \tag{2.5}$$

under which circumstances eqn. (2.4) simplifies to the normalized form

$$\begin{aligned}
\partial\theta_1/\partial\tau &= -\partial\theta_1/\partial h + \theta_2 - \theta_1 \\
\partial\theta_2/\partial\tau &= \partial\theta_2/\partial h - \theta_2 + \theta_1
\end{aligned} \tag{2.6}$$

where τ and h denote *normalized* time and distance, given by

$$\begin{aligned}
\tau &= t/T_n \\
h &= h'/L_n
\end{aligned} \tag{2.7}$$

where

$$\begin{aligned}
T_n &= 2S\rho A/(k\pi d W^{0.8}) \\
L_n &= 2SW^{0.2}/(k\pi d)
\end{aligned} \tag{2.8}$$

It is interesting to note that *base* time T_n is the time for either fluid to travel *base* distance L_n that in turn has an important physical significance which will become apparent later.

2.2.2 Binary distillation column

(*a*) *Packed type.* Distillation columns too are counterflow processes like the heat exchanger discussed above and can exhibit similarities in their behavioural characteristics. There are, however, important differences in their comparative behaviour resulting largely from the very different boundary conditions which apply in the two cases. These will be examined later but for the moment we shall concentrate on the development of the system's partial differential equations (p.d.e.'s) which will be found to resemble closely those of the heat exchanger. The system is illustrated by Fig. 2.2, which shows the two-stage construction of columns, involving two sections: the rectifier and the stripping section, and Fig. 2.3, which illustrates a conceptual cell of the rectifier, again assumed to be thoroughly mixed. Vapour and liquid streams flow past one another as shown at rates denoted by V and L moles† p.u. time, the streams being composed of mixtures of the two components to be separated by the column. X denotes the mole†-fraction (composition) of the more volatile (lighter) component in the liquid mixture and Y that in the vapour stream. Primes are associated with the variables in the stripping section and suffixes s and r associated with the flow rates denote stripping-section and rectifier quantities, respectively. As Fig. 2.2 shows, boiling mixture

---

† One mole of an element or compound has a weight numerically equal to the molecular weight expressed in the chosen system of units, e.g. one mole of water in SI units weighs $2 + 16 = 18$ kg.

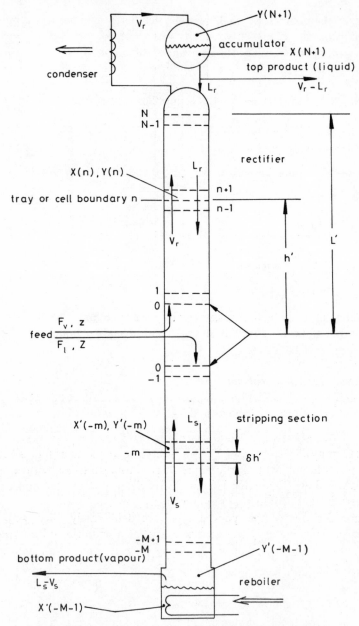

Fig. 2.2 *General arrangement of binary distillation column*

is fed into the column between stages, the liquid entering at flow rate F_l, composition Z and the vapour at flow rate F_v, composition z. Products are withdrawn from the process at top and bottom, i.e. from the accumulator and reboiler at rates $V_r - L_r$ and $L_s - V_s$ and at composition $X(N+1)$,

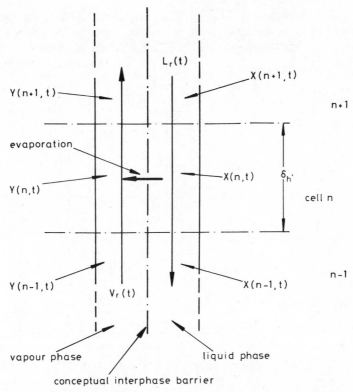

Fig. 2.3   *Variables associated with nth cell of column rectifying section*

$Y'(-M-1)$, respectively. The object of the distillation is to make $X(N+1)$ as close to unity as possible and $Y'(-M-1)$ close to zero at as high a throughput as possible and with a minimum energy utilization, i.e. with minimum recirculation flow. (Economic factors clearly determine the optimal compromise between these conflicting requirements.)

Column sections of the 'packed type' are deliberately filled with solid, pourous packing material to produce a spatial distribution of composition through the column so again necessitating consideration of an infinitesimal cell illustrated in Fig. 2.3. Within each cell, evaporation and condensation occur and under adiabatic conditions, if the two components have equal molecular latent heats, each mole condensing, of whatever component, causes another to evaporate. If only latent heats are considered therefore (sensible heat changes being assumed negligible by comparison), the heat balance for each cell is trivial and merely constrains the flow rates, V_r, V_s, L_r and L_s to be spatially invariant (as in the heat exchanger considered earlier but for different reasons).

In this process, unlike the heat exchanger, it is the material balance which produces the significant dynamic effects. If H_{rv}, H_{rl}, H_{sv} and H_{sl} denote the fixed molar capacitance p.u. length of the rectifier and stripping section for vapour and liquid respectively, then material balances for the lighter component taken on elementary slices of the two sections produce the following differential equations

$$
\left.
\begin{aligned}
H_{rv}\frac{dY}{dt}(n, t)\, \delta h' &= V_r\{Y(n - 1, t) - Y(n, t)\} \\
&\quad + k_r\{Y_e(n, t) - Y(n, t)\}\, \delta h' \\
H_{rl}\frac{dX}{dt}(n, t)\, \delta h' &= L_r\{X(n + 1, t) - X(n, t)\} \\
&\quad - k_r\{Y_e(n, t) - Y(n, t)\}\, \delta h'
\end{aligned}
\right\} \quad n > 1 \qquad (2.9)
$$

$$
\left.
\begin{aligned}
H_{sv}\frac{dY'}{dt}(n, t)\, \delta h' &= V_s\{Y'(n - 1, t) - Y'(n, t)\} \\
&\quad + k_s\{X'(n, t) - X'_e(n, t)\}\, \delta h' \\
H_{sl}\frac{dX'}{dt}(n, t)\, \delta h' &= L_s\{X'(n + 1, t) - X'(n, t)\} \\
&\quad - k_s\{X'(n, t) - X'_e(n, t)\}\, \delta h'
\end{aligned}
\right\} \quad n < -1 \qquad (2.10)
$$

where k_r and k_s are constant coefficients of evaporation and suffix e indicates *equilibrium* quantities. The terms involving these quantities (above) represent the net rate of evaporation of the lighter component, i.e. the cross-flow from the liquid to the vapour phase which ceases in situations where neighbouring liquid and vapour mixtures are in so-called 'thermodynamic equilibrium' with one another. For ideal mixtures (those obeying Dalton's and Rayoult's laws), the equilibrium relationship may be shown[1] to be

$$
\beta = Y_e(1 - X)/\{X(1 - Y_e)\} \qquad (2.11)
$$

where Y_e is the composition of a vapour in equilibrium with a liquid of composition X and, in terms of stripping section quantities:

$$
\beta = Y'(1 - X'_e)/\{X'_e(1 - Y')\} \qquad (2.12)
$$

where X'_e is the composition of a liquid mixture with which vapour of composition Y' would produce equilibrium. The parameter β is nearly constant for a given ideal mixture and is termed the 'relative volatility' of the mixture.

A typical equilibrium curve is sketched in Fig. 2.4 from which the symmetry about the $-45°$ line should be noted. β is greater than unity but the smaller its value the closer the curve approaches the $+45°$ line (i.e. the smaller the difference between the vapour and liquid compositions of equilibrium mixtures and the more difficult the distillation). For convenience of subsequent dynamic analysis, the curve is usually approximated[2] by two linear

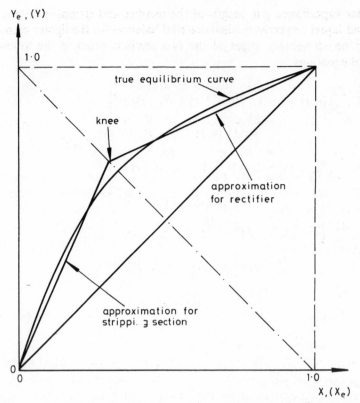

Fig. 2.4 *Ideal equilibrium curve for binary mixture and its piecewise-linear approximation*

relationships (one for the rectifier and the other for the stripping section), these being

$$\alpha(1 - Y_e) = 1 - X \qquad (2.13)$$

and

$$\alpha X'_e = Y' \qquad (2.14)$$

where the straight-line slopes are constrained thus:

$$\beta > \alpha > 1 \qquad (2.15)$$

so that, eliminating X and Y' using (2.13) and (2.14), we obtain

$$\left.\begin{array}{l}
H_{rv}\dfrac{dY}{dt}(n, t)\,\delta h' = V_r\{Y(n - 1, t) - Y(n, t)\} \\[2mm]
\qquad\qquad\qquad + k_r\{Y_e(n, t) - Y(n, t)\}\,\delta h' \\[4mm]
\alpha H_{rl}\dfrac{dY_e}{dt}(n, t)\,\delta h' = \alpha L_r\{Y_e(n + 1, t) - Y_e(n, t)\} \\[2mm]
\qquad\qquad\qquad - k_r\{Y_e(n, t) - Y(n, t)\}\,\delta h'
\end{array}\right\} \quad n > 1 \qquad (2.16)$$

and

$$H_{sl} \frac{dX'}{dt}(n, t)\, \delta h' = L_s\{X'(n+1, t) - X'(n, t)\}$$

$$\left.\begin{array}{r} - k_s\{X'(n, t) - X'_e(n, t)\} \\ \\ \alpha H_{sv} \dfrac{dX'_e}{dt}\, \delta h' = \alpha V_s\{X'_e(n-1, t) - X'_e(n, t) \\ \\ + k_s\{X'(n, t) - X'_e(n, t)\} \end{array}\right\} \; n < -1 \qquad (2.17)$$

We therefore have situations pertaining in the two column sections very similar in mathematical structure to those applying within the heat exchanger [cf. eqns. (2.1) and (2.2)] so that applying Taylor's theorem and letting $\delta h' \to 0$, as before, yields the p.d.e.'s

$$\left.\begin{array}{l} H_{rv}\, \partial Y/\partial t = -V_r\, \partial Y/\partial h' + k_r(Y_e - Y) \\ \alpha H_{rl}\, \partial Y_e/\partial t = \alpha L_r\, \partial Y_e/\partial h' - k_r(Y_e - Y) \end{array}\right\} \; h' > 0 \qquad (2.18)$$

$$\left.\begin{array}{l} H_{sl}\, \partial X'/\partial t = L_s\, \partial X'/\partial h' + k_s(X'_e - X') \\ \alpha H_{sv}\, \partial X'_e/\partial t = -\alpha V_s\, \partial X'_e/\partial h' - k_s(X'_e - X') \end{array}\right\} \; h' < 0 \qquad (2.19)$$

again very similar to those for the heat exchanger.

If we again confine attention to a symmetrical plant, i.e. one in which

$$\alpha H_{rl} = H_{sl} = H$$

$$H_{rv} = \alpha H_{sv} = cH \qquad (2.20)$$

$$k_r = k_s = k$$

where c is a constant, and operated under the nominal working conditions

$$V_r = \alpha L_r = L_s = \alpha V_s = V \qquad (2.21)$$

then normalizing the p.d.e.'s (2.18) and (2.19) yields the simplified system:

$$\left.\begin{array}{l} c\, \partial Y/\partial \tau = -\partial Y/\partial h + Y_e - Y \\ \partial Y_e/\partial \tau = \partial Y_e/\partial h - Y_e + Y \end{array}\right\} \; h > 0 \qquad (2.22)$$

$$\left.\begin{array}{l} \partial X'/\partial \tau = \partial X'/\partial h + X'_e - X' \\ c\, \partial X'_e/\partial \tau = -\partial X'_e/\partial h - X'_e + X' \end{array}\right\} \; h < 0 \qquad (2.23)$$

where τ and h denote *normalized* time and distance, being given by

$$\tau = t/T_n \qquad (2.24)$$

$$h = h'/L_n \qquad (2.25)$$

where

$$T_n = H/k \qquad (2.26)$$

and

$$L_n = V/k \tag{2.27}$$

Again the base distance/time ratio (L_n/T_n) = liquid velocity so that T_n is the time for the liquid to travel base distance L_n (the significance of which emerges later). The vapour/liquid capacitance ratio c will usually be $\ll 1.0$ but in later analysis will be set at unity in the interests of ease of solution.

(b) *Tray-type column*. Industrial scale column sections are more usually physically segmented by the deliberate inclusion of barriers or 'trays' holding constant volumes of liquid which cascades down the column from tray to tray. The vapour forces its way up through the trays by lifting so called 'bubble-caps' which act as non-return valves. With this construction a discretely changing spatial distribution of composition is achieved and our hitherto *conceptual* cells now acquire a definite physical significance. With this type of column it is generally assumed that the liquid and vapour above any given tray are in continuous equilibrium with one another and vapour capacitance is either neglected or lumped with the liquid capacitance. H_{rv} and H_{sv} are therefore put to zero in (2.9) and (2.10) and k_r and k_s made infinite so that

$$Y_e(n, t) = Y(n, t) \qquad n > 1$$

and

$$X'_e(n, t) = X(n, t) \qquad n < 1 \tag{2.28}$$

yielding general tray equations

$$\alpha H_{rl} \frac{dY}{dt}(n, t)\, \delta h' = \alpha L_r\{Y(n+1) - Y(n, t)\}$$
$$+ V_r\{Y(n-1, t) - Y(n, t)\} \qquad n > 1$$
$$H_{sl} \frac{dX'}{dt}(n, t)\, \delta h' = L_s\{X'(n+1) - X'(n, t)\}$$
$$+ \alpha V_s\{X'(n-1, t) - X'(n, t)\} \qquad n < -1 \tag{2.29}$$

$\delta h'$ here, of course, denotes the actual finite length of column between trays. Together with the boundary conditions, yet to be considered, numerical solution may therefore be undertaken at this stage with the advantage over the heat exchanger and packed column that the total number of cells (trays in this case) is prespecified. For analytical solution however, a p.d.e. representation is again preferable and is a permissible approximation when the column comprises a large number of trays, as is normally the case with industrial scale systems separating difficult mixtures. It is assumed that the discrete composition profiles may be closely approximated by spatially continuous functions so again permitting the use of the Taylor series to eliminate dependent variables other than $Y(n, t)$ and $X'(n, t)$, giving

$$\alpha H_{rl} \frac{\partial Y}{\partial t} = (\alpha L_r - V_r)\frac{\partial Y}{\partial h'} + \frac{(\alpha L_r + V_r)}{2}\frac{\partial^2 Y}{(\partial h')^2}\,\delta h' \qquad h' > 0$$

$$H_{sl} \frac{\partial X'}{\partial t} = (L_s - \alpha V_s)\frac{\partial X'}{\partial h'} + \frac{(L_s + \alpha V_s)}{2}\frac{\partial^2 X'}{(\partial h')^2}\,\delta h' \qquad h' < 0$$

(2.30)

ignoring higher powers of $\delta h'$.

Under the symmetrical operating conditions (2.20) and (2.21), the system therefore reduces to the normalized form

$$\partial Y/\partial \tau = \partial^2 Y/(\partial h)^2 \qquad h > 0$$

$$\partial X'/\partial \tau = \partial^2 X'/(\partial h)^2 \qquad h < 0$$

(2.31)

where again $\tau = t/T_n$ and $h = h'/L_n$, but the base time and distance are now given by

$$T_n = H\,\delta h'/V$$

$$L_n = \delta h'$$

(2.32)

The fundamental differences between the p.d.e.'s for packed and tray columns give rise to important differences in the dynamic behaviour as will be demonstrated later.

It should be emphasised that our treatment of columns has been confined to a consideration of composition dynamics. Columns are of course subject also to variations in internal pressure and in the levels of the end vessels, all of which interact with and are affected by the composition variations. The foregoing analysis has, however, assumed that these variables can be closely regulated, which is generally the case, but for a thorough investigation of these faster dynamics the reader is referred to the text of Rademaker *et al.*[3] Judson King[1] provides an excellent detailed coverage of steady-state column design.

2.2.3 Tubular chemical reactor

Having demonstrated similarities (and differences) between the mathematical structure of ideal heat- and mass-transfer processes, we now examine the influence of chemical change on process dynamics in situations where chemical kinetics dominate other factors. Mass transfer will be seen to play an important role easily embraced by the analysis. Thermal effects, though often crucial, are more difficult to include and temperatures will therefore be regarded as perfectly regulated in our investigation of the 'tubular' (spatially distributed) reactor. Uncontrolled temperature variations will, however, be examined afterwards in a consideration of the *continuous stirred-tank* reactor described by a lumped parameter model.

We consider the simple liquid reaction in which *reagents A* and *B* react together to form the single product *C*. For generality we will initially consider the reaction to be of the reversible type permitting *C* to decompose back into *A* and *B*. The *stoichiometric* equation for the reaction is therefore

$$A + B \rightleftarrows 2C \tag{2.33}$$

indicating that one mole of *A* reacts with one mole of *B* to form two moles of *C* and vice versa. (In practice the situation can be much more complicated involving gaseous and solid materials, intermediate products, several reactions taking place sequentially and in parallel, and the influence of catalysts etc., but the overall mechanism can often be constructed from a conceptual network of elementary systems of the sort examined here.)

Because of eqn. (2.33) we must now account for a new phenomenon in our basic concepts of dynamic balance: that of one type of material (mixture *A* and *B* in this case) changing into another type (here, product *C*). We therefore become involved with rates of chemical reaction which are found[4] from the kinetic theory of gases and experimentally, to be governed by equations of the type

$$r_c = k_1 [A]^\alpha [B]^\beta \tag{2.34}$$

and

$$r_{ab} = k_2 [C]^\gamma \tag{2.35}$$

where r_c denotes the rate of generation of *C* (and r_{ab} that of *A* and *B*) expressed in moles p.u. volume of mixture p.u. time, and the square brackets indicate *concentrations* of the appropriate substance expressed in moles per unit volume of the overall mixture. The *velocity coefficients* k_1 and k_2 are, for ideal gases, functions only of absolute temperature θ taking the form

$$k_1 = a_1 \exp\left(-\theta_{b1}/\theta\right) \qquad k_2 = a_2 \exp\left(-\theta_{b2}/\theta\right) \tag{2.36}$$

(where a_1, θ_{b1}, a_2 and θ_{b2} are constants) and are frequently assumed to be nearly so for reactions involving liquids and other non-ideal materials. The indices α, β and γ in eqns. (2.34) and (2.35) are usually small integers (or their reciprocals) generally determined experimentally.

Considering now the tubular reactor, a short section of which is illustrated in Fig. 2.5, then for conceptual cell *n* of length $\delta h'$ and volume δV, we may write down a material balance for any one of the three component substances *A*, *B* or *C*, taking account of the fact that components can now change from one to another, within the cell considered, at rates governed by equations of

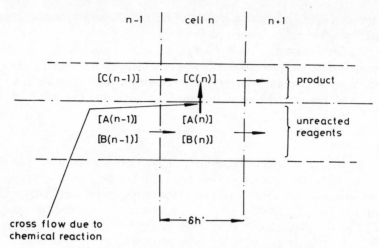

Fig. 2.5   *Conceptual model for tubular chemical reactor*

the type (2.34) and (2.35). Choosing C, the material balance for this substance may be written

$$\delta V \frac{d[C(n, t)]}{dt} = F\{[C(n-1, t)] - [C(n, t)]\}$$

$$+ \delta V\{k_1[A(n, t)]^{\alpha}[B(n, t)]^{\beta} - k_2[C(n, t)]^{\gamma}\} \qquad (2.37)$$

where F is the volumetric flow rate of mixture through the reactor.

Now if one mole of A, B or C occupy identical volumes, i.e. if the substances are *ideal*, then the molar density ρ_m of the overall mixture will be unchanged throughout, i.e.

$$\rho_m = [A(n, t)] + [B(n, t)] + [C(n, t)] = \text{a constant} \qquad (2.38)$$

and if reagents are fed in stoichiometric proportions (equal portions in this example) then

$$[A(n, t)] = [B(n, t)] \qquad (2.39)$$

so that

$$[A(n, t)] = [B(n, t)] = 0{\cdot}5\{\rho_m - [C(n, t)]\} \qquad (2.40)$$

permitting (2.37) to be expressed in terms of C only. The resulting differential equation becomes

$$\delta V \frac{d[C(n, t)]}{dt} = F\{[C(n-1, t)] - [C(n, t)]\}$$

$$+ \delta V\{k_1\rho_m/2 - (k_1/2 + k_2)[C(n, t)]\} \qquad (2.41)$$

if the reaction is of *first order* with respect to $[C]$, i.e. if

$$2\alpha = 2\beta = \gamma = 1\cdot 0 \qquad (2.42)$$

Alternatively, the composition of C could be expressed in terms of its mole fraction X rather than $[C]$, the number of moles p.u. volume, since

$$\rho_m X = [C] \qquad (2.43)$$

eqn. (2.41) thus becoming

$$\delta V \frac{dX(n, t)}{dt} = F\{X(n - 1, t) - X(n, t)\}$$

$$+ \delta V\{k_1/2 - (k_1/2 + k_2)X(n, t)\} \qquad (2.44)$$

from which we deduce, in a manner similar to that used for the previous examples, the p.d.e.

$$\partial X/\partial t = -(F/a)\,\partial X/\partial h' + (k_1/2 + k_2)\{k_1/(k_1 + 2k_2) - X\} \qquad (2.45)$$

where a is the cross-sectional area of the reactor tube.

Now $k_1/(k_2 + 2k_2)$ is the equilibrium value X_e which X would acquire were the reaction to take place in a *closed system*, such as a batch reactor, for a sufficient length of time. Under such circumstances backward and forward rates of reaction r_c and r_{ab} ultimately balance so that if suffix e denotes *equilibrium values*, then from eqns. (2.34) and (2.35) we get

$$k_2[C_e]^\gamma = k_1[A_e]^\alpha[B_e]^\beta \qquad (2.46)$$

which in our case reduces to

$$k_2[C_e] = k_1[A_e]^{0\cdot 5}[B_e]^{0\cdot 5} = k_1[A_e] \qquad (2.47)$$

if the initial charge of reagents were in stoichiometric proportion. Since, in general,

$$[A] + [B] + [C] = \rho_m \qquad (2.48)$$

and, in our case $[A] = [B]$, then

$$2[A] = \rho_m - [C] \qquad (2.49)$$

so that, substituting for $[A_e]$ in eqn. (2.47) we get

$$[C_e] = k_1\rho_m/(k_1 + 2k_2)$$

or

$$X_e = k_1/(k_1 + 2k_2) \qquad (2.50)$$

and so eqn. (2.45) becomes

$$\partial X/\partial t = -U\,\partial X/\partial h' + (k_1/2 + k_2)(X_e - X) \qquad (2.51)$$

where U is the fluid velocity. The equation may again be normalized giving a result identical, but for the symbols, to the previous spatially continuous examples, viz.

$$\partial X/\partial \tau = -\partial X/\partial h + X_e - X \qquad (2.52)$$

where $\tau = t/T_n$ and $h = h'/L_n$ and in this case

$$T_n = (k_1/2 + k_2)^{-1}$$
$$L_n = U(k_1/2 + k_2)^{-1} \qquad (2.53)$$

2.3 Similarities and differences in mathematical structure of the system models

In all three cases of the truly spatially continuous processes examined, the p.d.e.'s have involved only the first derivatives of the dependent variables with respect to space and time and cross-flow terms dependent upon the departure of these variables from their equilibrium values ($\theta_2(n, t)$ is the equilibrium value of $\theta_1(n, t)$ in the heat exchanger example and vice versa). Indeed similar p.d.e.'s are widely encountered in chemical plant modelling generally. It would therefore appear that any approach capable of producing a solution to one example should also cope with others provided the boundary conditions involved in the latter are of no greater complexity. We shall therefore focus attention on the distillation column to illustrate an approach to model solution since the process clearly involves the most complex boundary conditions of the examples considered.

Before proceeding, however, it is important to notice an important difference between the first two examples and the chemical reactor arising in connection with the multiplicative nature of the process eqns. (2.1) [and (2.2)], (2.18), (2.19) and (2.51). In the first two examples simple products occur, each involving one dependent variable (or its derivative) and one independent variable (or some power thereof): the flow rates in both cases being directly manipulable forcing functions. In the chemical reactor [eqn. (2.51)], however, $k_1/2 + k_2$ is a dependent variable unless tight temperature control can be exercised throughout the reactor length. In view of the great sensitivity of k_1 and k_2 to temperature θ [see eqns. (2.36)] and the often considerable rate at which heat is released or absorbed by reaction, the control of temperature can unfortunately pose a problem of greater magnitude than that of controlling composition X itself. This problem is therefore examined briefly before proceeding to the solution of the p.d.e.'s. The following section will also serve incidentally to introduce the concept of linearization in the fields of chemical process modelling.

2.4 Thermal characteristics of chemical reactors

The sometimes unusual nature of the thermal behaviour of reactors can be demonstrated by consideration of merely a single continuous stirred-tank reactor (CSTR) often employed in practice despite the less efficient utilization of the available volume compared to that achieved with a tubular reactor. The CSTR model is identical to that for a single cell representation of the tubular reactor and obtained by setting $n = 1$ in eqn. (2.41) and setting $[C(0, t)]$ to zero, assuming the reactor to be fed with reagents only. We shall also assume an irreversible reaction in this analysis so that k_2 is also zero. Hence, dropping the spatial argument n ($= 1$) of the variables we get

$$[\dot{C}(t)] = -(F/V)[C(t)] + \{k_1(\theta)/2\}\{\rho_m - [C(t)]\} \tag{2.54}$$

where the dot denotes the derivative with respect to time t, V, the total reactor volume, is set equal to δV and argument θ is associated with velocity k_1 to emphasize its temperature dependence. In terms of X rather than $[C]$, therefore, we obtain

$$\dot{X}(t) = -(F/V)X(t) + \{k_1(\theta)/2\}\{1 - X(t)\} \tag{2.55}$$

Now if θ is to vary, i.e.

$$\theta = \theta(t) \tag{2.56}$$

a heat balance is also necessary to complete the process model. We shall assume the reaction to be exothermic so that the rate of heat generation $q_g(t)$ by the reaction is given by

$$q_g(t) = \Delta H . r_c(t) . V \tag{2.57}$$

where ΔH is the energy released per mole of C produced. $r_c(t)$ is the rate of generation of C in moles p.u. volume and given by eqn. (2.34) so that using eqns. (2.40), (2.42) and (2.43) we get

$$q_g(t) = \Delta H . V . \rho_m k_1(\theta)\{1 - X(t)\}/2 \tag{2.58}$$

Now heat also enters the tank in the feed stream at a rate $F . H_c . \theta_a$ (if there is no preheating) and leaves in the outflow at a rate $F . H_c . \theta$, where θ_a is the ambient temperature and H_c the thermal capacitance p.u. volume of liquid. Heat will also be lost via the tank walls at a rate $Q_w(\theta - \theta_a)$† where Q_w is constant and heat may be deliberately extracted, by the immersion of, say, cooling tubes, at a rate $Q_t(\theta - \theta_c)$ where Q_t is an adjustable heat-transfer

---

† Assuming Newton's law of cooling to hold.

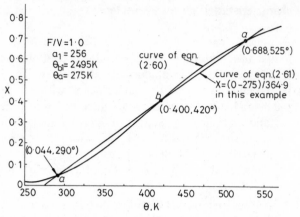

Fig. 2.6   *Showing possibility of triple solutions to steady-state reactor equations*

coefficient and θ_c the temperature of the cooling fluid, set for simplicity of analysis $= \theta_a$ in this example.† The heat balance thus becomes

$$H_c V \dot\theta(t) = \Delta H . V \rho_m k_1(\theta)\{1 - X(t)\}$$
$$- (Q_t + Q_w + FH_c).\{\theta(t) - \theta_a\} \qquad (2.59)$$

and thus, together with (2.55) and (2.36)—which determines the variation of k_1 with θ—provides a complete description of the thermally uncontrolled reactor. The steady-state equations of the process, obtained by setting $\dot X$ and $\dot\theta$ to zero, are

$$X = k_1(\theta)/\{k_1(\theta) + 2F/V\} \qquad (2.60)$$

and, eliminating $k_1(\theta)$ between the mass- and heat-balance equations,

$$X = \{(Q_t + Q_w)/F + H_c\}(\theta - \theta_a)/\Delta H . \rho_m \qquad (2.61)$$

the resulting curves of X versus θ being of sigmoid shape in the case of eqn. (2.60)‡ and a straight line in the case of eqn. (2.61) for constant Q_t and F. Depending upon the constant parameter values, therefore, single or triple points of intersection of the two curves are possible as illustrated in Fig. 2.6, suggesting the possibility of up to three steady-state working conditions.

2.4.1 Reactor linearization
The small-signal stability of the solutions may be investigated using the linearized, small-perturbation model of the system, derived by differentiating implicitly the non-linear large-signal d.e.'s (2.55) and (2.59) and setting the

---

† θ_c will, of course, vary to some extent with θ but will be nearly constant if a large flow of coolant is used.
‡ Since k_1 increases monotonically with increasing θ [see eqn. (2.36)].

differentials dX, $d\theta$, dF and dQ_t equal to small perturbations $x(t)$, $\phi(t)$, $f(t)$ and $q_t(t)$, respectively, giving

$$
\begin{bmatrix} \dot{x} \\ \\ \dot{\phi} \end{bmatrix} = \begin{bmatrix} -(F/V + k_1/2) & 0{\cdot}5(1 - X)\,\partial k_1/\partial\theta \\ & \Delta H . \rho_m(1 - X)(\partial k_1/\partial\theta)/2H_c \\ -\Delta H . \rho_m k_1/2H_c & -(Q_t + Q_w + H_c F)/H_c V \end{bmatrix} \begin{bmatrix} x \\ \\ \phi \end{bmatrix}
$$

$$
- \begin{bmatrix} X/V & 0 \\ \\ \dfrac{\theta - \theta_a}{V} & \dfrac{\theta - \theta_a}{H_c V} \end{bmatrix} \begin{bmatrix} f \\ \\ q_t \end{bmatrix} \tag{2.62}
$$

Provided the perturbations are sufficiently small in magnitude compared to the steady-state values of the variables X, θ, F and Q_t, then solutions of the large-signal steady-state eqns. (2.60) and (2.61) may be substituted as quasi-constants in the coefficient matrices of (2.62) so yielding the desired linearized system. If this system is written

$$
\begin{pmatrix} \dot{x} \\ \dot{\phi} \end{pmatrix} = \mathbf{A}\begin{pmatrix} x \\ \phi \end{pmatrix} + \mathbf{B}\begin{pmatrix} f \\ q_t \end{pmatrix} = \begin{pmatrix} a_{11} & a_{12} \\ a_{21} & a_{22} \end{pmatrix}\begin{pmatrix} x \\ \phi \end{pmatrix} + \begin{pmatrix} b_{11} & b_{12} \\ b_{21} & b_{22} \end{pmatrix}\begin{pmatrix} f \\ q_t \end{pmatrix} \tag{2.63}
$$

then the characteristic equation of the open-loop system is

$$
\text{Det } [\mathbf{I}s - \mathbf{A}] = 0 \tag{2.64}
$$

giving, in this case,

$$
s = \{(a_{11} + a_{22}) \pm \sqrt{(a_{11} - a_{22})^2 + 4a_{12}a_{21}}\}/2 \tag{2.65}
$$

so that for open-loop stability, i.e. $R_e s < 0$,

$$
(a_{11} - a_{22})^2 + 4a_{12}a_{21} < (a_{11} + a_{22})^2
$$

and

$$
a_{11} + a_{22} < 0
$$

These conditions may be more simply expressed:

$$
a_{11}a_{22} > a_{12}a_{21} \tag{2.66}
$$

and

$$
a_{22} < -a_{11} \tag{2.67}
$$

and substituting for the elements of $\mathbf{A}$ in (2.66), using (2.62) readily yields the *necessary* stability condition that

$$
(1 - X)\,\partial k_1/\partial\theta < 2(Q_t + Q_w + H_c F)(F/V + k_1/2)/F\,\Delta H . \rho_m \tag{2.68}
$$

while (2.67), after substitution, may be expressed

$$(1 - X)\, \partial k_1/\partial \theta < 2\{2H_c F(F/V) + H_c F k_1/2$$
$$+ (Q_t + Q_w)(F/V)\}/F\, \Delta H . \rho_m \qquad (2.69)$$

It is interesting to note that eqn. (2.68) has the immediate graphical interpretation that the slope of the sigmoid curve [eqn. (2.60)] in Fig. 2.6 should be less than that of the straight line [eqn. (2.61)] and Denbigh and Turner[4] offer a physical interpretation of the condition regarding the curves as heating and cooling characteristics, respectively. The arguments are not mathematically rigorous however (as the authors acknowledge), and the second condition (2.69) cannot be safely disregarded in general. The question is pursued more thoroughly by Himmelblau and Bischoff,[5] but necessary condition (2.68) does preclude intersections of type (*b*) shown in Fig. 2.6 from providing stable open-loop working points whereas intersections of type (*a*) might be stable or unstable. Clearly, the likelihood of contravening either condition is increased the larger ΔH becomes, i.e. the more exothermic the reaction, as would be expected.

This possibility of thermal runaway therefore poses a temperature control problem outside the scope of a text on process modelling but its analysis and solution is nevertheless crucial to the formulation of a model for studying the composition control problem.

2.5 Parametric transfer-function matrix models

We now return to the general problem of obtaining analytical solutions for the behaviour of chemical process plant. Using computer simulation techniques it is, of course, possible to proceed to numerical solutions directly from the system p.d.e.'s already derived, and the boundary conditions (yet to be considered), without further analysis, given:

(*a*) a reliable programmer
(*b*) values for the plant parameters and
(*c*) a control strategy for adjusting the manipulable input variables.

In practice, however, what is really required is a method for selecting the parameters and control strategy to produce a loosely prespecified mode of behaviour of the dependent variables. Posed this way round, computer solution of the problem is no longer direct since numerous iterations of the simulation will be required until the desired parameters and controller structure are, hopefully, determined. With the degrees of freedom possible, rapid convergence to the desired solution is not generally guaranteed.

An analytical model, in the form of a parametric transfer-function matrix (TFM), i.e. a TFM where parameters are known functions of plant parameters, is however directly usable by the plant designer and control engineers for the *synthesis*, in virtually one attempt, of the best plant/controller combination. The accurate derivation of such TFMs can be impractically tedious in the completely general case but if the plant possesses certain properties of linearity and symmetry (generally implied by good design as will be seen) such solutions can then be derived. Because of the discrepancies between the plant idealized in this way, and the true system model, simulation is still required, but now primed with a rationally determined initial controller structure and initial system parameters from which rapid convergence to the best solution, guided by the insight obtained from the analysed solution, might reasonably be expected.

Likewise, in the field of experimental identification of a difficult process model, analytical solution beforehand of the idealized system, does provide a soundly based model structure and good initial values for the iterative parameter-estimation exercise.

The procedure to be followed with spatially distributed systems of the type examined is broadly as follows, using the liquid/liquid heat exchanger to illustrate the steps involved. It is left to the reader to fill in some of the straightforward manipulations between steps.

2.5.1 Boundary condition formulation

The process description is incomplete without the boundary conditions which must first be determined. In the case of the heat exchanger these will be taken as being simply

$$\theta_1(0, t) = \text{constant} \tag{2.70}$$

$$\theta_2(L, t) = \text{constant} \tag{2.71}$$

where L is the *normalized* length of the process.

2.5.2 Large-signal steady-state solution

This is required to provide data for the quasi-constant parameters of the small-signal model and is obtained by setting time derivatives to zero in the process p.d.e.'s, (2.6) in this case, (and d.e.'s if any) and solving the resulting spatial d.e.'s subject to the boundary conditions (2.70) and (2.71). For the symmetrical heat exchanger the solution, graphed in Fig. 2.7, is

$$\theta_1(h) = \{\theta_1(0)(1 + L - h) + \theta_2(L)h\}/(L + 1)$$
$$\theta_2(h) = \{\theta_1(0)(L - h) + \theta_2(L)(1 + h)\}/(L + 1) \tag{2.72}$$

It will be noted from Fig. 2.7 that the two temperature profiles are separated by a normalized distance of 1·0, i.e. by an actual distance of L_n, the physical significance of which now emerges. The temperature profiles have equal constant gradients, which is a special result pertaining to the symmetrically operated process, producing a constant temperature drop $\theta_1(h) - \theta_2(h)$ so

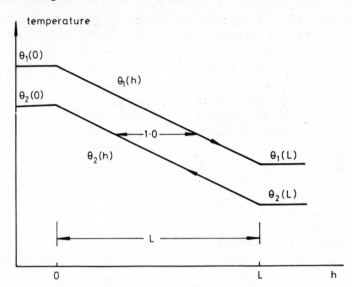

Fig. 2.7 *Steady-state temperature profiles for symmetrical heat exchanger*

that heat transfer is uniformly distributed along the interface, thereby making maximum use of the entire length of the process, i.e. good plant design.

2.5.3 Small-signal model derivation
As with the chemical CSTR the small-signal equations are derived from the large-signal equations by implicit differentiation, equating differentials to small perturbations in the dependent and independent variables, and substituting steady-state solutions for the coefficient values. With spatially distributed processes it is perhaps safer to operate on the large-signal equations before normalization since the base time, T_n, and base distance, L_n, may be functions of the variables. The small-signal equations may then be converted to their simpler normalized form afterwards. From large-signal p.d.e.'s (2.4) we thus obtain

$$\partial\phi_1/\partial\tau = -\partial\phi_1/\partial h + \phi_2 - \phi_1 + f_1(\tau)$$
$$\partial\phi_2/\partial\tau = \partial\phi_2/\partial h + \phi_1 - \phi_2 + f_2(\tau) \tag{2.73}$$

where flow functions f_1 and f_2 are given by

$$\begin{bmatrix} f_1 \\ f_2 \end{bmatrix} = \frac{\theta_1(0) - \theta_2(L)}{W(L+1)} \begin{bmatrix} 0\cdot6, & -0\cdot4 \\ 0\cdot4, & -0\cdot6 \end{bmatrix} \begin{bmatrix} w_1 \\ w_2 \end{bmatrix} \tag{2.74}$$

(Had the steady-state gradients $\partial\theta_1/\partial h$ and $\partial\theta_2/\partial h$ not been constant then forcing functions f_1 and f_2, which involve these gradients, would have been spatially dependent as well as time varying, so complicating subsequent analysis considerably.) The small-signal boundary equations derived from eqns. (2.70) and (2.71) in this example are simply

$$\phi_1(0, \tau) = 0 \qquad \phi_2(L, \tau) = 0 \tag{2.75}$$

2.5.4 Laplace transformation with respect to h and τ

Taking Laplace transforms of the small-signal p.d.e.'s first in p w.r.t. τ and then in s w.r.t. h yields, in this case,

$$p\tilde{\tilde{\phi}}_1(s, p) = -(1 + s)\tilde{\tilde{\phi}}_1(s, p) + \tilde{\tilde{\phi}}_2(s, p) + \tilde{\phi}_1(0, p) + s^{-1}\tilde{f}_1(p)$$
$$p\tilde{\tilde{\phi}}_2(s, p) = -(1 - s)\tilde{\tilde{\phi}}_2(s, p) + \tilde{\tilde{\phi}}_1(s, p) - \tilde{\phi}_2(0, p) + s^{-1}\tilde{f}_2(p) \tag{2.76}$$

(for zero initial conditions), in which $\approx$ denotes transforms w.r.t. h and τ and $\sim$ w.r.t. τ only. Boundary conditions specified at $h = 0$ can be eliminated at this stage, here by simply putting $\tilde{\phi}_1(0, p)$ to zero as demanded by eqn. (2.75) but $\tilde{\phi}_2(0, p)$ is presently unknown and must therefore be retained until after inversion back to the space domain.

2.5.5 Inversion to the h, p domain

Having isolated separate expressions from eqn. (2.76) for $\tilde{\tilde{\phi}}_1(s, p)$ and $\tilde{\tilde{\phi}}_2(s, p)$ in terms of inputs $\tilde{f}_1(p)$, $\tilde{f}_2(p)$ and the unknown $\tilde{\phi}_2(0, p)$ these may now be inverted from the s, p to the h, p domain with the aid of Laplace transform tables and, by substituting $L = h$, the second boundary condition of eqn. (2.75) may be invoked to yield an expression for $\tilde{\phi}_2(0, p)$ which may then be used to determine $\tilde{\phi}_1(L, p)$. These two results grouped into matrix form are conveniently expressed thus:

$$\begin{bmatrix} \tilde{\phi}_1(L, p) - \tilde{\phi}_2(0, p) \\ \tilde{\phi}_1(L, p) + \tilde{\phi}_2(0, p) \end{bmatrix} = \begin{vmatrix} \theta_1(0) - \theta_2(L) \\ W(L + 1) \end{vmatrix} G(p) \begin{bmatrix} \tilde{w}_1(p) + \tilde{w}_2(p) \\ \tilde{w}_1(p) - \tilde{w}_2(p) \end{bmatrix} \tag{2.77}$$

where the TFM $G(p)$ takes the diagonal form

$$\mathbf{G}(p) = \begin{bmatrix} g_1(p) & 0 \\ 0 & g_2(p) \end{bmatrix} \tag{2.78}$$

where

$$g_1(p) = \frac{0.2\{p(\cosh qL - 1) + q \sinh qL\}}{q\{q \cosh qL + (1 + p) \sinh qL\}} \tag{2.79}$$

and

$$g_2(p) = \frac{(q^2/p)(\cosh qL - 1) + q \sinh qL}{q\{q \cosh qL + (1 + p) \sinh qL\}} \tag{2.80}$$

the frequency function q being given by

$$q^2 = p(p + 2) \tag{2.81}$$

Alternatively, in terms of the real-life input and output vectors, rather than their so-called 'tilt' and 'total' combinations, (2.77) may be expressed

$$\begin{bmatrix} \tilde{\phi}_1(L, p) \\ \tilde{\phi}_2(0, p) \end{bmatrix} = \frac{|\theta_1(0) - \theta_2(L)|}{|2W(L + 1)|} \begin{bmatrix} 1 & 1 \\ -1 & 1 \end{bmatrix} \begin{bmatrix} g_1(p) & 0 \\ 0 & g_2(p) \end{bmatrix}$$
$$\times \begin{bmatrix} 1 & 1 \\ 1 & -1 \end{bmatrix} \begin{bmatrix} \tilde{w}_1(p) \\ \tilde{w}_2(p) \end{bmatrix} \tag{2.82}$$

the TFM between this input and output vector having a 'dyadic' structure,[6] because the dynamics of the system are contained entirely within a diagonal matrix that is coupled to the observed outputs and manipulable inputs by purely static matrices. This structure arises from the physical symmetry of the process considered and occurs frequently in analytically derived TFM models because tractable analytic solutions are generally limited to symmetrical cases—as has been already emphasized. Since control system design lies outside the scope of this text, suffice it to note that the dyadic nature of system's TFM clearly makes the solution of the interaction problem in controller design a fairly trivial exercise exhaustively investigated by Owens[6] (who has also considered the application of dyadic approximation which would be applicable to plants operated with a degree of asymmetry). The problem therefore reduces essentially to the design of stable controllers for the individual diagonal terms of $G(p)$, the computed inverse Nyquist loci for which are shown in Figs. 2.8 and 2.9 for the case of $L = 2·0$. These are clearly directly usable for control system synthesis.

Approximations to $g_1(p)$ and $g_2(p)$ may also be derived (a) by simplification of their accurate formulae (2.79) and (2.80) to provide a check on the computation of the true loci or (b) directly from the small-signal p.d.e.'s and boundary conditions to avoid the labour of deriving precise solutions altogether. The approach resembles in some respects those of Owens[7] and Friedly[8] and involves the matching of asymptotic models derived for very high and very low ranges of frequency.

2.5.6 Multivariable first-order lag models
Owens[7] has proposed that if a system has an inverse TFM $G^*(p)$ where

$$\lim_{p \to \infty} p^{-1}G^*(p) = A_0 \tag{2.83}$$

and

$$\lim_{p \to 0} G^*(p) = A_1 \tag{2.84}$$

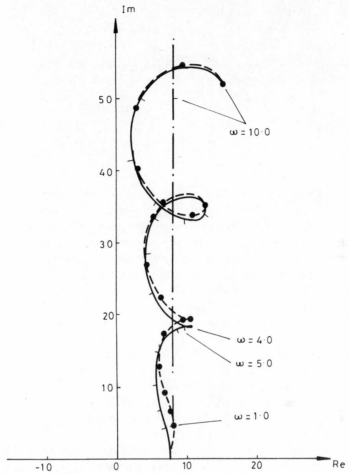

Fig. 2.8 *Locus of $g_1^{-1}(j\omega)$*
———— true system locus
– – – – – lag/delay approximation
—·—·— m.v. 1st-order lag approximation ω-increment – 0.5 except where indicated otherwise

where A_0 and A_1 are constant matrices then, under certain conditions, $G^*(p)$ may be approximated, for controller design purposes, by a multivariable first-order lag system of inverse TFM $G_A^*(p)$ where

$$G_A^*(p) = A_1 + A_0 p \tag{2.85}$$

since $G^*(p)$ and $G_A^*(p)$ approach equality at very high and very low frequencies. Such a representation is highly appropriate and very convenient in chemical plant modelling since A_1 and A_0 are readily determined either from

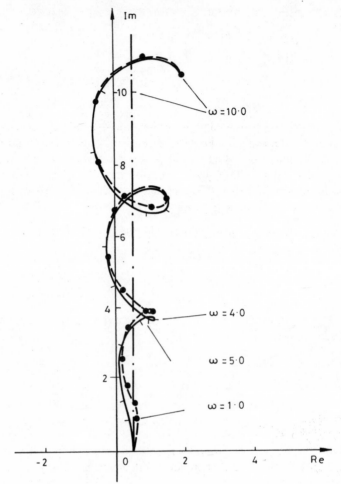

Fig. 2.9 *Locus of* g_2^{-1} *(jω)*

——— true system locus

– – – – lag/delay approximation

—·—·— m.v. 1st-order lag approximation ω-increment = 0.5 except where indicated otherwise

measurements of the initial rates and the settling values of the system's step responses or by simple analytical derivation. From the transformed system p.d.e.'s (2.76) in this example $\mathbf{A}_1$ is readily deduced by merely ignoring all dependent variables other than those with coefficients involving the highest power of p. In this case clearly we get

$$\lim_{p \to \infty} p \begin{bmatrix} \tilde{\phi}_1(h, p) \\ \tilde{\phi}_2(h, p) \end{bmatrix} = \begin{bmatrix} \tilde{f}_1(p) \\ \tilde{f}_2(p) \end{bmatrix} \tag{2.86}$$

from which, transforming to our tilt and total variables we get

$$\underset{p \to \infty}{\text{Lim}} \, p \begin{bmatrix} \tilde{\phi}_1(L, p) - \tilde{\phi}_2(0, p) \\ \tilde{\phi}_1(L, p) + \tilde{\phi}_2(0, p) \end{bmatrix} = \left| \frac{\theta_1(0) - \theta_2(L)}{W(L + 1)} \right|$$
$$\times \begin{bmatrix} 0.2 & 0 \\ 0 & 1 \end{bmatrix} \begin{bmatrix} \tilde{w}_1(p) + \tilde{w}_2(p) \\ \tilde{w}_1(p) - \tilde{w}_2(p) \end{bmatrix} \quad (2.87)$$

from which matrix A_0 is immediately obtainable. It is interesting to note that A_0 is spatially independent and independent of boundary conditions upon which A_1 is crucially dependent. A_1 is in fact determined by the solution of the small-signal p.d.e.'s (2.73) with constant inputs subject to the system boundary conditions (2.75) in this example with time derivatives set to zero. In the case of the heat exchanger this yields the result

$$\begin{bmatrix} \phi_1(L) - \phi_2(0) \\ \phi_1(L) + \phi_2(0) \end{bmatrix} = \left| \frac{\theta_1(0) - \theta_2(L)}{W(L + 1)} \right| \begin{bmatrix} 0.2L/(L + 1) & 0 \\ 0 & L \end{bmatrix}$$
$$\times \begin{bmatrix} w_1 + w_2 \\ w_1 - w_2 \end{bmatrix} \quad (2.88)$$

From eqns. (2.87) and (2.88) we therefore deduce the multivariable first-order lag model for the system to be

$$\begin{bmatrix} \tilde{\phi}_1(L, p) - \tilde{\phi}_2(0, p) \\ \tilde{\phi}_1(L, p) + \tilde{\phi}_2(0, p) \end{bmatrix} \simeq \frac{\theta_1(0) - \theta_2(L)}{W(L + 1)}$$
$$\times \begin{bmatrix} 0.2\{(L + 1)/L + p\}^{-1} & 0 \\ 0 & (1/L + p)^{-1} \end{bmatrix} \begin{bmatrix} \tilde{w}_1(p) + \tilde{w}_2(p) \\ \tilde{w}_1(p) - \tilde{w}_2(p) \end{bmatrix} \quad (2.89)$$

the inverse Nyquist loci for which are also shown in Figs. 2.8 and 2.9, for $L = 2.0$, alongside the true system loci. Agreement is clearly good but the loops in the true loci could produce unexpected oscillation or even instability in the presence of high-gain integral control action.

2.5.7 Lag/delay models

Failure to predict the loops in the inverse Nyquist loci arises from neglecting the imaginary nature of p when considering the high-frequency asymptotic behaviour of the system. If, therefore, only the interactive terms (rather than all dependent variables not multiplied by p) are omitted from eqn. (2.76), since $\tilde{\phi}_1(0, p) = 0$ we obtain,

$$\tilde{\phi}_1(s, p) = f_1(p)/s(s + p + 1)$$

and

$$\tilde{\phi}_2(s, p) = \{\tilde{\phi}_2(0, p) - \tilde{f}_2(p)/s\}/\{s - (p + 1) \quad (2.90)$$

giving, on inversion and substituting $h = L$, [since $\bar{\phi}_2(L, p) = 0$],

$$\begin{bmatrix} \bar{\phi}_1(L, p) \\ \bar{\phi}_2(0, p) \end{bmatrix} = \frac{1 - \exp\{-(p + 1)L\}}{(p + 1)} \begin{bmatrix} \tilde{f}_1(p) \\ \tilde{f}_2(p) \end{bmatrix} \qquad (2.91)$$

Now substituting for $\tilde{f}_1(p)$ and $\tilde{f}_2(p)$ in terms of $\tilde{w}_1(p)$ and $\tilde{w}_2(p)$ we obtain the high-frequency model:

$$\operatorname*{Lim}_{|p| \to \infty} \left(\frac{p}{1 - \exp\{-(p + 1)L\}} \right) \begin{pmatrix} \bar{\phi}_1(L, p) - \bar{\phi}_2(0, p) \\ \bar{\phi}_1(L, p) + \bar{\phi}_2(0, p) \end{pmatrix}$$

$$= \frac{|\theta_1(0) - \theta_2(L)|}{|W(L + 1)|} \begin{pmatrix} 0\cdot2 & 0 \\ 0 & 1\cdot0 \end{pmatrix} \begin{pmatrix} \tilde{w}_1(p) + \tilde{w}_2(p) \\ \tilde{w}_1(p) - \tilde{w}_2(p) \end{pmatrix} \qquad (2.92)$$

which resembles the high-frequency eqn. (2.87) apart from the appearance of the attenuated delay-term $\exp\{-(p + 1)L\}$ and the indication that the limit applies irrespective of p being real or complex. The same result may be derived from the accurate model (2.77) to (2.81) noting that, as $|p| \to \infty$,

$$q \to p + 1\cdot0 \qquad (2.93)$$

Combining the inverse system TFM's for low and high frequency in a manner similar to that for first-order lag approximation we therefore now obtain the multivariable lag/delay model:

$$\begin{pmatrix} \bar{\phi}_1(L, p) - \bar{\phi}_2(0, p) \\ \bar{\phi}_1(L, p) + \bar{\phi}_2(0, p) \end{pmatrix} \simeq \frac{\theta_1(0) - \theta_2(L)}{W(L + 1)}$$

$$\times \begin{bmatrix} 0\cdot2\left(\dfrac{(L + 1)}{L} + \dfrac{p}{1 - \exp\{-(p + 1)L\}}\right)^{-1} & 0 \\ 0 & \left(\dfrac{1}{L} + \dfrac{p}{1 - \exp\{-(p + 1)L\}}\right)^{-1} \end{bmatrix} \begin{pmatrix} w_1(p) + w_2(p) \\ w_1(p) - w_2(p) \end{pmatrix} \qquad (2.94)$$

which exhibits loops in its inverse Nyquist loci very similar to those of the real system as inspection of Figs. 2.8 and 2.9 reveals.

The loops are in fact the result of reflected travelling waves in the process which become progressively more attenuated with the passage of time. The first-order lag model predicts only the fundamental integrating nature of the system at high frequency whilst the lag-delay model reproduces the effect of the passage of the first of these waves in addition.

The analytical approach to T.F.M. development outlined in this section is next applied to the distillation process described earlier. This is a much more difficult case because of (a) its two-stage construction, therefore involving four boundaries not two, and (b) the relative complexity of the individual boundary equations. Solution is, however, not impossibly tedious if full advantage is taken of physical symmetry to simplify the system equations at every step of the analysis. Because of space limitations only the key intermediate results are provided here but the reader should not have great difficulty in performing the intermediate manipulations. The fully worked-out analyses are to be found in References 9, 10 and 11.

2.6 The analytical determination of parametric TFM's for symmetrical distillation columns

2.6.1 *Large-signal boundary conditions*
Without these the system description is incomplete and their formulation requires the consideration of the mass balances pertaining at the top and bottom of both the rectifier and stripping sections. Their final forms differ somewhat in the cases of packed and tray-type columns largely because of the fixed, finite cell length in the latter case. Considering the feed point, firstly for packed columns, it is readily deduced that, for the vapour and liquid streams, respectively:

$$V_s Y'(0) + F_v z = V_r Y(0)$$
$$L_r X(0) + F_l Z = L_s X'(0)$$

(2.95)

and it follows from the assumed symmetry conditions (2.21) that the liquid and vapour feed flows must be equal and given by

$$F_v = F_l \triangleq F = V\epsilon/\alpha$$

(2.96)

where

$$\epsilon = \alpha - 1$$

(2.97)

For symmetry we shall also assume that the feed mixture is supplied in equilibrium *for both sections* so that

$$z = \alpha Z \quad \text{and} \quad z = 1 - Z$$

(2.98)

so fixing z and Z to fixed nominal working values. For the tray column however considering the tray above the feed point we deduce that

$$H_l \, \delta h' \, dX(0)/dt = Fz + V_s Y'(0) - V_r Y(0) + L_r\{X(1) - X(0)\}$$

(2.99)

so that approximating the finite difference in X by the first spatial derivative, eliminating X and Y' in favour of Y and X', substituting for z and normalizing we get

$$\partial Y(0)/\partial \tau = -2/(\alpha + 1) + X'(0) + \{1 - Y(0)\} + \partial Y(0)/\partial h$$

(2.100)

and similarly for the tray beneath the feed point

$$\partial X'(0)/\partial \tau = 2(\alpha + 1) - X'(0) - \{1 - Y(0)\} - \partial X'(0)/\partial h$$

(2.101)

N.l.—E

Turning attention now to the top (accumulator) end of the rectifier ($h = L$), we have, for the packed column

$$H_e \, dX(L)/dt = V_r Y(L) - V_r X(L) \tag{2.102}$$

where H_e is the capacitance of the accumulator. Assuming this vessel to run in continuous equilibrium therefore

$$H_e \alpha \, dY_e(L)/dt = V_r[\alpha\{1 - Y_e(L)\} - \{1 - Y(L)\}] \tag{2.103}$$

Similar consideration applied to the reboiler ($h = -L$) gives

$$H_e \alpha \, dX'_e(L)/dt = L_s[X'(-L) - \alpha X'_e(-L)] \tag{2.104}$$

if αH_e is the reboiler capacitance. The tray column's terminal conditions are similar to their finite difference form, i.e. for the accumulator:

$$H_e \, dX(N + 1)/dt = V_r\{Y(N) - X(N + 1)\} \tag{2.105}$$

but

$$X(N + 1) \simeq X(N) + \{\partial X(N)/\partial h'\} \, \delta h' \tag{2.106}$$

so that eliminating X in favour of Y using the equilibrium relationship (2.13) and normalizing yields

$$T_e \, \partial Y(L)/\partial \tau = \epsilon\{1 - Y(L)\} - \partial Y(L)/\partial h \tag{2.107}$$

and similarly for the reboiler

$$T_e \, \partial X'(-L)/\partial \tau = -\epsilon X'(-L) + \partial X'(-L)/\partial h \tag{2.108}$$

where T_e is the normalized time constant of the end vessels given by

$$T_e = H_e/H \, \delta h' \tag{2.109}$$

2.6.2 Large-signal steady-state solution

With the formulation of the four boundary conditions the two columns are now completely specified and steady-state solutions for constant inputs (V_r, L_r, F_v, F_l, Z and z) may be determined by the simultaneous solution of p.d.e.'s (2.22), (2.23) with (2.95), (2.103) and (2.104) for the packed column, and p.d.e.'s (2.31) with (2.100), (2.101), (2.107) and (2.108) for the tray column, putting the time derivatives to zero beforehand. Because of the symmetry of the system equations resulting from operating at flow conditions (2.2.1) and with input compositions governed by eqn. (2.98), the composition profiles $Y(h)$, $Y_e(h)$, $X'(h)$ and $X'_e(h)$ are readily determined and are found to be linear. The solutions are graphed in Fig. 2.10 in which G is the composition gradient given by

$$G = 2\epsilon/(\alpha + 1)(2\epsilon L + \alpha + 1) \tag{2.110}$$

The equal separation per unit length of column is again characteristic of a

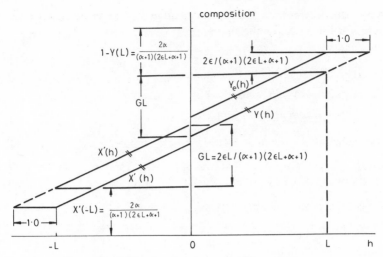

Fig. 2.10 *Steady-state composition profiles*

well-designed plant making full use of the available column volume. Solutions $Y(h)$ and $X'(h)$ for the tray column are found to be identical to those for $Y_e(h)$ and $X'_e(h)$ in the packed column because of the continuous equilibrium assumption. The equality of $X'(-h)$ and $1 - Y(h)$ and of $X'_e(-h)$ and $1 - Y_e(h)$ is an important symmetrical property of these profiles which also greatly eases solution of the small-signal model now to be considered.

2.6.3 Small signal p.d.e.'s and Laplace transformation

2.6.3.1 Reversal of rectifier distance base. It will be recalled from the heat exchanger case that the first dependent variable to emerge from the solution was $\bar{\phi}_2(0, p)$ from which $\bar{\phi}_1(L, p)$ and indeed $\bar{\phi}_1(h, p)$ and $\bar{\phi}_2(h, p)$ could then be obtained by substitution of the $\bar{\phi}_2(0, p)$ expression. The important dependent variables in the column are the output perturbations $y(L)$ and $x'(-L)$ in $Y(L)$ and $X'(L)$ and labour is therefore saved if these two emerge first from the analysis, so saving additional substitution of perhaps complex expressions. The distance base is therefore now altered, replacing h by $L - h$ in the rectifier equations and h by $L + h$ in the stripper equations, so that $h = L$ now locates the feed point in both cases and $h = 0$ locates the top *and* bottom of the entire column. $y(L)$ and $x'(-L)$ in the original system of co-ordinates therefore now become $y(0)$ and $x'(0)$ whilst $y(0)$ and $x'(0)$ in the original base now become $y(L)$ and $x'(L)$. The transformation is best understood by imagining the column bent into an inverted U-tube with the feed point now at the top of the U and both the reboiler *and* accumulator now at the bottom. The transformation clearly reverses the sign of the odd but not the even spatial derivatives in the rectifier equations. The new base is assumed in all subsequent equations.

By implicit differentiation of the general packed column p.d.e.'s (2.18) and (2.19) and substitution of the symmetrical steady-state solutions we get, after normalizing,

$$c \, \partial y/\partial \tau - \partial y/\partial h + Gv/V = y_e - y$$
$$-\partial y_e/\partial \tau - \partial y_e/\partial h + \alpha Gl/V = y_e - y$$
$$-\partial x'/\partial \tau + \partial x'/\partial h + Gl/V = x' - x'_e$$
$$c \, \partial x'_e/\partial \tau + \partial x'_e/\partial h + \alpha Gv/V = x' - x'_e$$

(2.111)

in which v and l denote perturbations in inputs V_r and L_r, whilst the boundary conditions (2.45), (2.103) and (2.104) yield (in the new base)

$$y(L) = x'_e(L) - (\epsilon/2)Gv/V$$
$$x'(L) = y_e(L) + (\epsilon/2)Gl/V$$

(2.112)

$$\alpha(1 + T \, \partial/\partial\tau)y_e(0) = y(0)$$
$$\alpha(1 + T \, \partial/\partial\tau)x'_e(0) = x'(0)$$

(2.113)

Taking Laplace transforms of the p.d.e.'s in s w.r.t. h and in p w.r.t. τ produces

$$(1 + cp - s)\tilde{y} - \tilde{y}_e + G\tilde{v}/Vs + \tilde{y}(0) = 0$$
$$-(1 + p + s)\tilde{y}_e + \tilde{y} + \alpha G\tilde{l}/Vs + \tilde{y}_e(0) = 0$$
$$-(1 + p - s)\tilde{x}' + \tilde{x}'_e + G\tilde{l}/Vs - \tilde{x}'(0) = 0$$
$$(1 + cp + s)\tilde{x}'_e - \tilde{x}' + \alpha G\tilde{v}/Vs - \tilde{x}'_e(0) = 0$$

(2.114)

whilst transforming eqn. (2.113) w.r.t. τ only gives

$$\tilde{y}_e(0) = \alpha^{-1}h_e(p)\tilde{y}(0)$$
$$\tilde{x}_e(0) = \alpha^{-1}h_e(p)x'(0)$$

(2.115)

where $h_e(p)$ is the transfer function of the end vessels, i.e.

$$h_e(p) = 1/(1 + Tp)$$

(2.116)

The system clearly becomes completely symmetrical if we set c, the column's vapour/liquid capacitance ratio, $= 1{\cdot}0$† producing both dynamic as well as static symmetry.

2.6.4 Matrix representation

Adding and subtracting the analogous equations of set (2.114) produces equations in composition *tilts* and *totals* resembling their temperature equivalents in Section 2.5. Furthermore identical coefficients appear in the resulting equa-

---

†This assumption implies high-pressure distillation. It is not essential to the tractability of the solution but is helpful inasmuch as it produces diagonal coefficient matrices.

tions so yielding matrix equations with purely diagonal coefficient matrices which greatly assists solution. In particular, if input and output vectors are defined thus

$$\mathbf{q} = \begin{pmatrix} y - x' \\ y + x' \end{pmatrix} \qquad \mathbf{r} = \begin{pmatrix} y_e - x'_e \\ y_e + x'_e \end{pmatrix} \quad \text{and} \quad \mathbf{u} = \frac{G}{V}\begin{pmatrix} v + l \\ v - l \end{pmatrix} \qquad (2.117)$$

we get $(1 + p - s)\tilde{\mathbf{q}} - \tilde{\mathbf{r}} = -s^{-1}\tilde{\mathbf{u}} - \tilde{\mathbf{q}}(0)$ and

$$-(1 + p + s)\tilde{\mathbf{r}} + \tilde{\mathbf{q}} = \alpha s^{-1}\begin{pmatrix} -1, 0 \\ 0, 1 \end{pmatrix}\tilde{\mathbf{u}} - \tilde{\mathbf{r}}(0) \qquad (2.118)$$

whilst similar operations on the boundary conditions (2.112) and (2.115) produce:

$$\mathbf{q}(L) = \begin{pmatrix} -1, 0 \\ 0, 1 \end{pmatrix}\mathbf{r}(L) - \frac{\epsilon}{2}\mathbf{u} \qquad (2.119)$$

and

$$\tilde{\mathbf{r}}(0) = \alpha^{-1}h_e(p)\tilde{\mathbf{q}}(0) \qquad (2.120)$$

2.6.5 Inversion to the h, p domain

Again the boundary conditions at $h = 0$ [eqn. (2.120)] may be used immediately to eliminate say the unknown $\tilde{\mathbf{r}}(0)$ from the transformed p.d.e.'s which may then be manipulated and inverted before substituting the second pair of boundary conditions for $h = L$ into eqn. (2.119) so yielding the desired solution for $\tilde{\mathbf{q}}(0, p)$, and hence for $y(0, p)$ and $x'(0, p)$. This is found to be

$$\tilde{\mathbf{q}}(0, p) = \begin{pmatrix} g_1(0, p) & 0 \\ 0 & g_2(0, p) \end{pmatrix}\tilde{\mathbf{u}}(p) \qquad (2.121)$$

where

$$g_1(0, p) = \frac{\{\epsilon(\cosh qL - 1)/p - (1 + \alpha)[\sinh qL]/q - \epsilon/2\}}{\{(1 - h_e\alpha^{-1})[\sinh qL]q/p + (1 + h_e\alpha^{-1})(\cosh qL)\}} \qquad (2.122)$$

and

$$g_2(0, p) = \frac{\{\epsilon p(\cosh qL - 1)/q^2 - (\alpha + 1)[\sinh qL]/q - \epsilon/2\}}{\{(1 + h_e\alpha^{-1})[\sinh qL]p/q + (1 - h_e\alpha^{-1})(\cosh qL)\}} \qquad (2.123)$$

where again

$$q^2 = p(p + 2) \qquad (2.124)$$

and, for zero frequency, i.e. step inputs

$$g_1(0, 0) = \alpha\{\epsilon L^2 - (\alpha + 1)L - \epsilon/2\}/\{2\epsilon L + \alpha + 1\} \qquad (2.125)$$

and

$$g_2(0, 0) = -\alpha\{(\alpha + 1)L + \epsilon/2\}/\epsilon \qquad (2.126)$$

Of course these results apply to the packed column, but an analysis of the tray column may be carried out on lines† similar to those demonstrated in Sections 2.6.3. to 2.6.5, producing a result as eqn. (2.121) but in which the elements are here given by slightly different formulae, viz.

$$g_1(0, p)$$

$$= \frac{\epsilon\{(p + 2)(\cosh \sqrt{p}L - 1)/p + (\sinh \sqrt{p}L)/\sqrt{p} + 0\cdot5\}}{\{(1 + T)p + 2 + \epsilon\} \cosh \sqrt{p}L + \{(p + 2)(\epsilon + Tp) + p\}(\sinh \sqrt{p}L)/\sqrt{p}}$$

$$(2.127)$$

$$g_2(0, p)$$

$$= -\frac{(\alpha + 1)(\cosh \sqrt{p}L - 1) + (\alpha + 1) \sinh \sqrt{p}L/\sqrt{p} + 0\cdot5(3\alpha + 1)}{\{p(1 + T) + \epsilon\} \cosh \sqrt{p}L + \sqrt{p}(1 + \epsilon + Tp) \sinh \sqrt{p}L}$$

$$(2.128)$$

$$g_1(0, 0) = (\epsilon L^2 + \alpha + 1)/(2\epsilon L + \alpha + 1) \qquad (2.129)$$

and

$$g_2(0, 0) = -\{(\alpha + 1)L + 0\cdot5(3\alpha + 1)\}/\epsilon \qquad (2.130)$$

2.6.6 Form of the inverse Nyquist loci

Two examples, for packed columns, of the behaviour $g_1(0, j\omega)^{-1}$ are illustrated in Fig. 2.11 for the following parameters

(a) $\epsilon = 0\cdot75$, $(\alpha = 1\cdot75)$, $L = 2\cdot8$ $T = 20$ and
(b) $\epsilon = 1\cdot0$, $(\alpha = 2\cdot0)$ $L = 5\cdot0$ $T = 20$

In both cases loops due to travelling-wave effects are clearly visible and obviously of greater significance in the case of the shorter column (a) which, as might be expected, causes less attenuation of the waves between reflections. It is also very important, however, to note that the sign of the static gain $g_1(0, 0)$ is parameter-dependent, being positive for longer columns where

$$\epsilon L^2 > (\alpha + 1)L + \epsilon/2 \qquad (2.131)$$

whereas, over higher-frequency ranges, the gain is invariably negative. Larger columns of the packed type therefore produce non-minimum-phase open-loop behaviour (and therefore severe closed-loop stability constraints) as indicated by the encirclement of the origin by locus (b), which would not be predicted by, say, first-order lag modelling, based on eqn. (2.85) due to the opposing signs of corresponding elements of matrices A_i and A_0.

---

† Although vector **r** as defined in (2.117) will not appear in this case, the double spatial derivative of **q** will also generate *two* unknown vectors at $h = 0$ upon transforming the system p.d.e.'s, so again requiring two sets of boundary conditions.

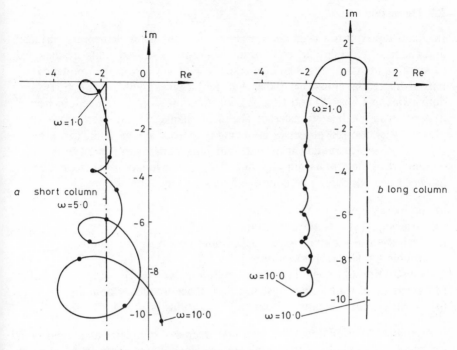

Fig. 2.11   *Loci of g_1^{-1} (0, jω) for packed column*
——————— true system loci
—·—·—·— m.v. 1st order lag approximations ω-increment = 1.0

[The true loci are compared in Fig. 2.11 with their multivariable first-order lag approximations derived from eqn. (2.125) and the approximate high-frequency model

$$\lim_{p \to \infty} p\tilde{\mathbf{q}} = -\tilde{\mathbf{u}} \tag{2.132}$$

obtained from the first equation of (2.118)].

First-order lag modelling is, however, applicable to tray columns,[12, 13] the high- and low-frequency gains of which are of identical sign† (positive for $g_1(0, p)$ and negative for $g_2(0, p)$). Travelling-wave effects are here found to be unimportant because the validity of the spatially continuous model for tray columns depends upon L being ≫ 1. In the case of shorter packed columns, i.e. those not satisfying eqn. (2.131), first-order lag modelling may also be applied, but here the additional phase lag caused by the significant wave effects must also be included for high-gain controller design.

———————————————
† Rosenbrock[14] in 1966 first indicated the possibility of important behavioural differences between packed and tray type columns resulting from the basic differences in their p.d.e.'s. It has nevertheless taken until now for these differences to be identified thoroughly[9, 10, 11, 15] probably because of the avoidance of the analytical approach by the majority of researchers.

2.7 Discussion

In this Chapter, it has been demonstrated that, by using elementary physical and chemical balance and equilibrium concepts, idealized units of chemical process plant, involving significant spatial variation, can be modelled by partial differential equations (p.d.e.'s) of similar type, involving only first-order derivatives in space (h) and time (τ), whether the dominant phenomenon is material transfer, heat transfer or chemical change. It has been shown that physically discretized processes involving numerous stirred tanks can also be represented approximately by p.d.e.'s involving second-order spatial derivatives. The analytical determination of parametric transfer-function matrices (TFMs) has been demonstrated involving the following sequence of steps

(*a*) boundary condition formulation
(*b*) large signal steady-state solution
(*c*) derivation of the small-signal p.d.e.'s and boundary conditions
(*d*) double Laplace transformation
(*e*) substitution of known boundary conditions (at $h = 0$)
(*f*) inversion of transformed system to the space-frequency domain
(*g*) substitution of remaining boundary conditions (at $h = L$).

Analytical solution does demand a high degree of symmetry and linearity in the plant equations but fortunately this is also a characteristic of good plant design. The symmetry leads to TFMs of dyadic structure, the control design for which is much more straightforward than for multivariable systems in general because of the ease with which interaction can be removed.

The examples considered have revealed that the fundamental high-frequency behaviour is dictated by the p.d.e.'s alone whereas boundary conditions dominate low-frequency behaviour. Because of the very diverse range of boundary conditions possible from process to process therefore, very different overall dynamic behaviour can be expected from processes governed by similar p.d.e.'s including non-minimum-phase responses. Travelling-wave phenomena can be important in processes of length insufficient to cause significant attenuation between wave reflections and in such circumstances lag/delay models can closely reproduce true system behaviour. In other circumstances multivariable first-order lag approximations provide rapid approximate solutions provided high- and low-frequency gains are of identical sign, i.e. provided the system is of a minimum-phase type.

This chapter has been restricted to processes involving one dominant physical or chemical phenomenon though it has been demonstrated that chemical/thermal interactions can in practice demand the consideration of these two effects simultaneously. The analytical solution of the idealized decomposed system can nevertheless provide, in general, good initial parameter values and controller strategies with which to begin detailed computer or pilot-plant simulations.

2.8 References

1 JUDSON KING, C.: 'Separation Processes' (McGraw-Hill, New York, 1971)
2 WILKINSON, W. L., and ARMSTRONG, W. D.: *Chem. Eng. Sci.*, 1957, **7**(1/2)
3 RADEMAKER, O., RIJNSDORP, J. E., and MAARLEVELD, A.: 'Dynamics and control of continuous distillation units' (Elsevier, Amsterdam, 1975)
4 DENBIGH, K., and TURNER, J. C. R.: 'Chemical reactor theory' (Cambridge University Press, 1973)
5 HIMMELBLAU, D. M., and BISCHOFF, K. B.: 'Process analysis and simulation— Deterministic systems' (J. Wiley and Son, New York, 1968)
6 OWENS, D. H.: 'Dyadic expansions and their applications', *Proc. IEE*, 1979, **126**(6), pp. 563–567
7 OWENS, D. H.: 'Feedback and multivariable systems', IEE Control Engineering Series 7 (Peter Peregrinus, London, 1978)
8 FRIEDLY, J. C.: 'Asymptotic approximations to plug flow process dynamics', *AIChE*, June 1967
9 EDWARDS, J. B.: 'The analytical modelling and dynamic behaviour of a spatially continuous binary distillation column', University of Sheffield, Dept. of Control Engineering, Research Report No. 86, April 1979
10 EDWARDS, J. B.: 'The analytical modelling and dynamic behaviour of tray-type binary distillation columns', *ibid.*, Research Report No. 90, June 1979
11 EDWARDS, J. B.: 'The analytical determination of the composition dynamics of binary distillation columns of the packed- and tray-type', *Trans. Inst. Chem. Eng. (GB)*, to be published
12 EDWARDS, J. B., and JASSIM, H. J.: 'An analytical study of the dynamics of binary distillation columns', *ibid.*, 1977, **55**, pp. 17–28
13 EDWARDS, J. B., and OWENS, D. H.: 'First-order type models for multi-variable process control', *Proc. IEE*, 1977, **124**(11), pp. 1083–1088
14 ROSENBROCK, H. H., and STOREY, C.: 'Computational techniques for Chemical Engineers' (Pergamon Press, London, 1966)
15 EDWARDS, J. B.: 'The dynamic behaviour of packed and tray-type binary distillation columns described by lumped parameter models', University of Sheffield, Dept. of Control Engineering, Research Report No. 91, June 1979

Chapter 3

Refrigeration and air conditioning systems

R. W. James

3.1 Introduction

The physical events taking place within refrigeration and air-conditioning systems are normally described mathematically by sets of algebraic equations which are derived by applying the steady flow energy equation to each section of the system. An American Society of Heating, Refrigeration and Air-Conditioning Engineers Task Group on energy requirements for heating and cooling recommended in 1971 procedures for simulating components and system performance with steady-state models. In practice, steady-state conditions are rarely obtained.

In order to design systems which will fulfil their functions adequately whilst minimizing energy consumption and capital cost, a knowledge of the transient behaviour of refrigeration and air-conditioning systems, and the conditioned enclosures is required. For this reason, mathematical models to simulate a variety of buildings have been developed, in addition progress is being made in the modelling of refrigeration and air-conditioning systems.

In this chapter the derivation and use of mathematical models giving transient and steady-state responses of refrigeration and air-conditioning systems are discussed. Although the author would regard the building or refrigerated enclosure as part of the system, they are not discussed in detail. The thermal behaviour of buildings is very complex and the reader can refer to the references given for information on mathematical models giving their transient behaviour. Information is given on how the reader can increase his understanding of how buildings behave using simple dynamic models giving limited information on particular aspects of buildings, and the simulation of a double brick wall is given as an example.

3.2 Modelling of refrigeration systems

Mathematical models giving transient and steady-state responses of single and two-stage refrigeration systems have been derived using the 'stirred-tank' approach. To derive such a model it is necessary to consider each section of a plant in turn, the sections can then be zoned and equations can be written to represent each zone. Extensive sorting and manipulation of the equations is normally necessary before the model is finally ready for programming. A typical refrigeration system incorporating a compressor, condenser, evaporator and suction/liquid line heat exchanger is shown in Fig. 3.1.

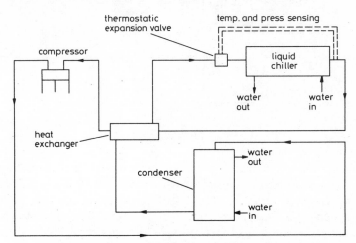

Fig. 3.1 *A typical refrigeration system*

3.2.1 Evaporators

The behaviour of refrigerants inside evaporator tubes is complex but can broadly be categorized in terms of 'flow regimes', which describe the characteristic distribution of the fluid-to-fluid interface in a two-phase fluid-flow system. Many types of two-phase flow exist, each with a unique range of flow regimes. Figure 3.2 shows the flow regimes that occur in refrigeration condensers and evaporators. Prediction of the flow regime is difficult, often flow maps similar to the one proposed by Baker[1] (Fig. 3.3) are used, but no reliable method has yet been found. In refrigeration system evaporators, annular flow is by far the most common flow regime encountered.

Until more reliable information is available enabling mathematical models to be based on the actual flow regime the following method previously used by Marshall[2] for steam generators has been shown to give reasonable results[3]. The liquid chiller shown in Fig. 3.4 can be zoned as shown by dividing the unit into sections to represent the distributive behaviour of the liquid surrounding the tubes, the refrigerant boiling inside the tubes and the tubes themselves.

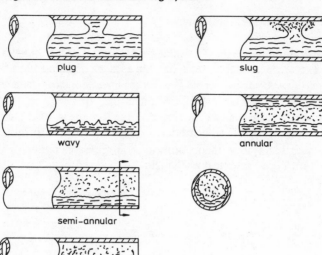

Fig. 3.2   *Two-phase fluid flow regimes*

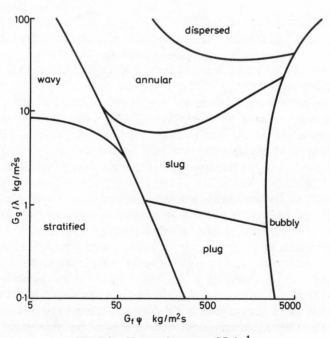

Fig. 3.3   *Flow regime map of Baker[1]*

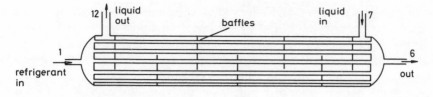

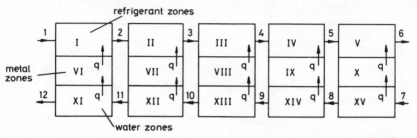

Fig. 3.4   *The inner-fin dry expansion liquid chiller (evaporator) and assumed zoning*

The author uses five zones for each making a total of fifteen zones, but reasonable results have been achieved with only three zones representing the refrigerant, three for the liquid being cooled, and increasing the volume of the liquid to allow for the thermal inertia of the metal without representing the metal with separate zones, so reducing the total number of zones to six. If this simplification is made with air coolers, experience has shown that the responses obtained will be a poor representation of the system.

Equations can be derived from energy and mass balance on each zone in turn, which, for the refrigerant in zone I, Fig. 3.4, gives

$$m_1 h_1 + q_{1,2} - m_2 h_2 = V_I \frac{d}{dt}(\rho_2 h_2) \tag{3.1}$$

$$m_1 - m_2 = V_I \, d\rho_2/dt \tag{3.2}$$

for the water in zone XI

$$m_{11} h_{11} - q_{11,12} - m_{12} h_{12} = V_{XI} \rho_{12} \, dh_{12}/dt \tag{3.3}$$

$$m_{11} - m_{12} = 0 \tag{3.4}$$

and for the metal

$$q_{11,12} - q_{1,2} = MCp \, dT_{VI}/dt \tag{3.5}$$

where m is the mass flow rate in kg/s, h is the enthalpy in kJ/kg, q is the heat transfer in kW, V is the volume in m^3 and ρ is the density in kg/m^3. For eqns. (3.3) and (3.4) it is assumed that the density of liquid does not change significantly.

Algebraic equations must be derived for each zone so that there are as many equations as variables in the model. State equations giving the enthalpy, density and temperature of saturated liquid and vapour as functions of pressure are derived by plotting graphs from property tables and fitting polynomials. Second-order polynomials give good results, or linear equations can be used if the operating range is small, e.g. the following equation gives good results for refrigerant 12 (CF_2Cl_2) in the temperature range 0–10°C

$$h_f = 0.077p + 12.6 \tag{3.6}$$

This equation gives the enthalpy of saturated liquid as a function of pressure. If a state equation is required to represent superheated vapour the variable will be a function of pressure and temperature, for example, for R12 in the same temperature range, density is given by

$$\rho = p[0.057 - 0.000175(T - T_g)] \tag{3.7}$$

Density and enthalpy are normally calculated from differential equations similar to (3.1) and (3.2). Equation (3.7) would be combined with the following equations which are for the same temperature range, and manipulated to give an equation for pressure in terms of enthalpy and density only

$$T = T_g + 1.667(h - h_g) \tag{3.8}$$

$$h_g = 0.033p + 177.6 \tag{3.9}$$

$$T_g = 0.091p - 28.2 \tag{3.10}$$

A series solution can then be obtained with the equations for the variables that are functions of pressure following the derived equation for pressure.

The refrigerant mass flow rate can be calculated by using momentum equations relating force and flow rate. For zone *II* (Fig. 3.4), this will be

$$m_2 = [(p_2 - p_3)\rho_2/C]^{1/2} \tag{3.11}$$

where C is a constant depending on the friction factor, tube geometry etc., and is best determined experimentally wherever possible. It is important to use the density at the zone inlet in eqn. (3.10), otherwise stability of the model will not be achieved.

Equations giving the dryness fraction at any section can be derived in terms of the enthalpy, and the enthalpy of saturated liquid and vapour, i.e.

$$x = (h - h_f)/h_{fg} \tag{3.12}$$

For the liquid being cooled, temperature is a function of enthalpy, and the temperature of the metal is obtained from a differential equation similar to eqn. (3.5). The heat-transfer equations can therefore be written provided that the heat-transfer coefficients are known. Normally they are a function of the fluid mass flow rate but when the operating temperature range is large and the fluid properties change significantly, they must also be functions of temperature. For zone I the equation is as follows

$$q_{1,2} = A\alpha(T_{VI} - T_{g12}) \tag{3.13}$$

The accuracy of the model can be improved by using a log mean temperature difference, although a step change of inlet temperature would immediately change $q_{1,2}$. However, this does not occur in the evaporator being modelled, therefore this refinement is only valid when rapid changes in inlet temperature cannot occur.

To represent the evaporator of Fig. 3.4, 21 differential equations and 56 algebraic equations were used together with various inputs, including the volume of each zone, density of water, heat-transfer coefficients and surface areas. For liquid chillers, the differential equations giving the density of the refrigerant have fast responses, and these can be replaced with algebraic equations which give a significant saving on computer time, although for a series solution the combined equations introduce implicit equations into the model.

These equations are readily solved using the Newton-Raphson technique, but the model now gives less information about events taking place in the evaporator.

For air coolers with large external fins, the fin effectiveness must be taken into account. A small reduction of the area used in the heat-transfer equations is sufficient to do this and it is normally unnecessary to use more than one metal zone for each refrigerant zone.

3.2.2 Condensers

The range of flow regimes normally encountered in refrigeration condensers is greater than those encountered in evaporators, however the method used by Marshall[2] for evaporators gives reasonable results. With an ammonia industrial refrigeration system, the simple conceptual model of an evaporative condenser shown in Fig. 3.5 gives surprisingly good results, but this was mainly because the model was intended to cover summer operation only[3].

The conceptual model comprises a vapour space (I), a boundary layer (II), a liquid space (III) and the metal (IV). The vapour space represented the volume of the tubing between the compressor and condenser, the condenser tubes and the vapour space in the liquid receiver above the liquid level. The boundary layer was assumed to have no capacity and as a result of heat transfer the vapour is condensed in this section and once liquid it was assumed to be in section III.

An energy balance on each section in turn gives equations similar to (3.1) and (3.2). The total volume of the actual liquid receiver, condenser and delivery tube was $2\cdot4$ m^3, hence

$$V_v + V_l = 2\cdot4 \tag{3.14}$$

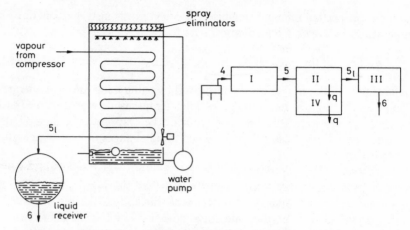

Fig. 3.5 *Evaporative condenser and liquid receiver with the assumed zoning.*

From a momentum equation relating force and flow rate

$$p_4 = p_5 + 630 \cdot 0 m_5^2 / \rho_5 \tag{3.15}$$

where

$$m_5 = q_5 / (h_5 - h_{f5})$$

To allow for the heat capacity of the metal an equation similar to eqn. (3.5) was derived. The heat-transfer equations were similar to eqn. (3.13), but the temperature difference used to calculate q was assumed to be that between the metal and wet bulb temperature of the surrounding air. This would not have been satisfactory for winter operation when air cooling inside evaporative condensers represents a much greater proportion of the total heat transfer taking place. The state equations were derived in the manner discussed for evaporators.

3.2.3 Shell and coil condensers
Figure 3.6 shows the shell and coil condenser together with the 'conceptual model' which summarizes the assumed zoning of the condenser system. Superheated vapour enters the vessel at connection 2 and is then represented by section I in the conceptual model. A proportion of the vapour in this section enters the thermal boundary layer and is cooled; once cooled it is less buoyant and, depending on its position, it either moves away from the tube wall or condenses against it. The conceptual model consists of the following zones:

(*a*) a vapour space I, to represent the vapour outside the boundary layer
(*b*) a boundary layer II, containing both liquid and vapour
(*c*) a return link (4) for the vapour escaping from the boundary layer,
(*d*) a liquid space III, for the condensed refrigerant

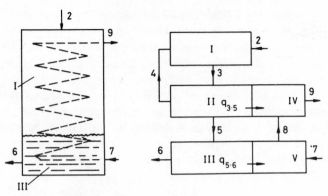

Fig. 3.6 *Shell and coil condenser and the conceptual model.*
　I Vapour outside boundary layer
　II Boundary layer
　III Liquid refrigerant
　IV Water in coil above refrigerant liquid level
　V Water in coil below refrigerant liquid level

(*e*) a space V to represent the water in the coil below the R12 liquid level
(*f*) a space IV to represent the water in the coil above the R12 liquid level.

The heat transfer that occurs between the refrigerant and the condenser cooling water is represented by a heat transfer between sections II and IV above the R12 liquid level and between sections III and V below in the conceptual model.

An energy and mass balance on each of the sections in Fig. 3.6 yields equations similar to (3.1) and (3.2). Since the volume of the boundary layer is small compared to other volumes in the model the differential equations for this section would act fast and they were therefore replaced by algebraic equations by setting the differential coefficients to zero.

The mass flow of vapour entering the boundary layer and subsequently escaping from it via connection 4 in the conceptual model is difficult to determine. However, the ratio of the mass of vapour escaping to that condensed can be assumed to be constant, which yields the following equation

$$m_4 = Cm_5 \qquad\qquad (3.16)$$

The constant can be programmed as an input and should be corrected by making a comparison between the temperature of the vapour inside the vapour space on the plant and model, although the temperature of the vapour inside the condenser is difficult to measure. The value of C was assumed to be approximately equal to 7.

The state equations are derived as for the evaporator and the density of liquid refrigerant was assumed to be constant.

The total volume of the vessel is constant, and an equation similar to (3.14) can be derived for the refrigerant and for the water in the cooling coil. It is important to note that for the refrigerant

$$\frac{dV_l}{dt} = -\frac{dV_v}{dt} \tag{3.17}$$

where V_l and V_v is the volume of liquid and vapour respectively. A similar equation can be derived for the water in the cooling coil.

Equations similar to (3.13) can be derived for the heat transfer between the refrigerant and tube, and between the tube and water both above and below the refrigerant liquid level. The surface areas in the heat-transfer equations are functions of the refrigerant liquid level.

3.2.4 Shell and tube condensers

In shell and tube condensers the shell is of cylindrical construction with its axis horizontal and the tubes containing the water pass horizontally through the vessel from end to end. They can be treated in the same way as shell and coil condensers. If there is a separate liquid receiver, the vapour volume in the condenser can be considered to be constant, however if the bottom of the condenser is used as a liquid receiver, the surface area of the tubes containing water above and below the refrigerant liquid surface will not be a linear function of liquid level. The situation becomes even more complex if the water passes through the tubes in an order dictated by ease of manufacture rather than through the lower ones, then working steadily upwards to the top tubes. The equations derived must take this into account.

3.2.5 Compressors

Events that take place inside the cylinders of reciprocating compressors have been modelled by MacLaren,[4] who had a specific interest in valve behaviour. However, the equations derived would respond much faster than other system equations and unless the compressor itself is of specific interest it is best represented by algebraic equations.

A close approximation to the enthalpy increase across a compressor is made by assuming that it is equal to the indicated power, thus

$$\Delta h = K_1(T + 273)[(r)^{K_2} + 1] \tag{3.18}$$

where r is the pressure ratio and K_1 and K_2 are constants that can be determined by assuming reasonable values for the specific heat and index of compression.

An alternative is to use the following equation to calculate the compressor discharge temperature

$$\frac{T_2}{T_1} = \left(\frac{p_2}{p_1}\right)^{(n-1)/n} \tag{3.19}$$

where T is the absolute temperature and n the index of compression which is normally above the value of the isentropic index. For R12 the index is usually about 1·2.

To calculate the mass flow rate the volumetric efficiency must be known. Normally, manufacturers of compressors will provide graphs of volumetric efficiency plotted against the pressure ratio for a range of operating speeds and equations that can be fitted to these graphs. State equations for the refrigerant at the compressor inlet and exit can again be obtained by plotting and fitting. An equation of the form

$$h = h_g + Cp(T - T_g) \tag{3.20}$$

can be used to relate enthalpy and temperature of the superheated vapour, and a mean value of the specific heat can be assumed since it varies with both pressure and temperature.

The final equation giving the mass flow rate will be as follows:

$$m = K\rho\eta_v \tag{3.21}$$

where K is a constant depending on the swept volume of the cylinders, number of cylinders and compressor operating speed. The inlet and outlet mass flow rates can be assumed to be equal.

3.2.6 Suction/liquid line heat exchangers
The suction/liquid line heat exchanger is used to undercool the liquid leaving the condenser using the cold vapour leaving the evaporator. The total heat transfer in such a unit is normally small and the assumed zoning can consist of one zone for the liquid and one for the vapour. The state and heat-transfer equations can be derived as for the evaporator section and the heat-transfer coefficients can be found experimentally or from manufacturers' data, or calculated from dimensionless equations in the normal way.

3.2.7 Thermostatic expansion valves
If the refrigerant flow rate into the evaporator is monitored by a thermostatic expansion valve sensing the pressure and temperature of the refrigerant at evaporator outlet, the mass flow rate is a function of the difference between the refrigerant vapour temperature and its saturation temperature at the evaporator outlet. The thermostatic expansion valve can normally be represented as a one-term proportional controller.

Figure 3.7 shows the block diagram[5] for a thermostatic expansion valve which has to maintain the desired superheat of the vapour leaving the evaporator.

The position of the needle is the result of the interaction of three forces on the flexible diaphragm shown, i.e.

F_1—which is exerted as a result of the pressure of the liquid/vapour mixture in the remote phial

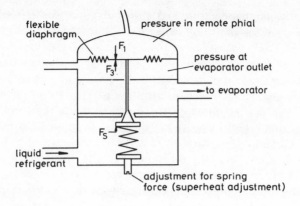

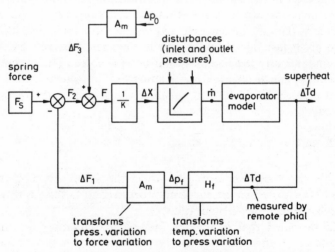

Fig. 3.7 *The thermostatic expansion valve and the block diagram representing its dynamics*

F_3—which is the result of the pressure of the refrigerant at evaporator outlet
F_s—which is the force exerted by the spring.

The remote phial normally contains the same refrigerant as the plant itself and therefore F_1 and F_3 are equal if liquid refrigerant reaches the position of the sensor at evaporator outlet. The force exerted by the spring is adjustable and has to be overcome by the difference between F_1 and F_3 before the valve begins to open; the difference between these forces is a function of suction superheat at evaporator outlet and therefore the adjustment is known as 'the superheat adjustment'.

The resultant of F_s and F_1 gives a variation in the force F_2 in the block diagram, and the resultant of F_2 and F_3 gives a variation F in the force which is transmitted to the regulating circuit as a signal. Through the block representing the spring characteristic $1/K$ this signal is transformed into a variation in the valve lift X. The variation in the lift X mm is transformed through the block for the valve characteristic into a variation in the amount of refrigerant injected into the evaporator.

In the block for the evaporator, $\dot{m}$ is transformed into a new variation in the superheat T_d at the evaporator outlet, which causes a repetition of the sequence of the regulating process described until a steady-state condition is achieved. In the block for the valve characteristic, variations in condensing and evaporating pressures and the difference between them act as disturbances, since a variation in the pressure drop across the valve results in a variation in the amount of refrigerant injected into the evaporator at the same lift.

It is not difficult to represent these processes mathematically but if there is no specific interest in the stability of the evaporator-expansion valve control loop, approximations can be made. Pressure variations can be assumed to produce force variations immediately and displacement of the needle can be assumed to occur fast. The delay is then in the signal source which cannot change instantaneously because heat transfer to or from the remote phial (sensing element) has to take place. Assuming that like most sensors, a first-order lag is involved, the following equation will represent the thermostatic expansion valve

$$\frac{dm}{dt} = -m + \text{GAIN}(T - T_g - T_{\text{sup}}) \qquad 0 < m < m_{\text{max}} \qquad (3.22)$$

where T_{sup} is the superheat setting at which the valve will start to open. Equation (3.22) assumes a time constant of one second but for most valves it would be substantially higher than this.

It should be noted that if large variations in evaporator or condenser pressure take place, this equation will be inadequate to represent the dynamics of the thermostatic expansion valve. The GAIN and m_{max} can of course be made functions of the pressure drop across the valve.

3.2.8 Intercoolers for two-stage systems
The combined intercooler and liquid subcooler shown in Fig. 3.8 is used widely in low-temperature two-stage industrial refrigeration systems. The refrigerant inside this vessel is at the intermediate pressure and the vapour from the first-stage compressor cylinders is bubbled through the liquid refrigerant inside this vessel and is thus cooled. Liquid refrigerant from the receiver is circulated through the subcooling coil which is immersed in the liquid refrigerant inside the intercooler. The subcooled liquid refrigerant leaving this vessel is now delivered to the evaporators and as a result of its temperature being reduced, less vapour is formed as it passes across the expansion valve from the high to the low-pressure region.

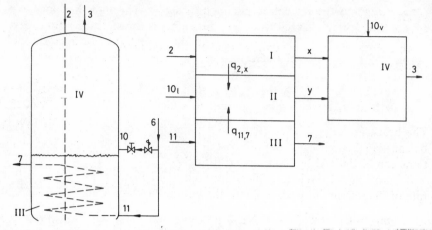

Fig. 3.8 *Combined intercooler and liquid subcooler for a two-stage refrigeration system and the conceptual model*
  I Vapour bubbles from first stage compressor cylinders
  II Liquid refrigerant
  III Sub-cooling coil
  IV Vapour space

The conceptual model shown in Fig. 3.8 can be used to represent the dynamics of the actual combined intercooler and liquid subcooler. The four sections shown represent the vapour bubbles below the surface of the liquid, the liquid refrigerant inside the vessel, the subcooling coil and the vapour space above the surface of the liquid.

When the liquid level inside this vessel falls to a preset value, refrigerant enters the vessel at connection 10 and in passing from the high to the intermediate-pressure region part of the feed will evaporate. In the model, the feed that remains as liquid enters section II whilst the vapour formed in passing through the expansion valve enters section IV.

The mass flow rate m_y is the result of the liquid refrigerant in section II evaporating. This occurs because (a) heat is added to the liquid refrigerant inside the intercooler by the subcooler coil, and (b) the highly superheated vapour from the first stage compressor cylinders is bubbled through it. This combined effect is represented in the conceptual model by the mass flow rate m_y from section II to IV. When the vapour bubbles, which are present as a result of the superheated vapour entering the vessel from the first-stage compressor cylinders, reach the surface of the liquid, they pass from section I to IV and this is represented by the mass flow rate m_x on the conceptual model.

The heat transfer that takes place between the highly superheated vapour entering the intercooler and the liquid inside this vessel is represented on the conceptual model by q_{2x} which crosses the boundary between sections I and II. Likewise the heat transfer that takes place between the liquid refrigerant inside the subcooling coil and the liquid surrounding it ($q_{11,7}$) crosses the boundary between sections III and II.

An energy and mass balance is made on all sections, for example, an energy balance on the liquid space (section II) gives:

$$q_{11,7} + q_{2x} + m_{10l}h_{10l} - m_y h_y = \frac{d}{dt}(V_l \rho_l h_l) \tag{3.23}$$

The heat-transfer coefficients that are required to calculate $q_{11,7}$ and q_{2x} can be calculated empirically, but are best determined experimentally. State equations can be derived by plotting and fitting.

If on-off liquid level controllers are used, the equations for the mass flow rate of the refrigerant feed, m_{10} into the liquid separator are

$$m_{10} = C \quad \text{if} \quad V_l < K_1$$
$$m_{10} = 0 \quad \text{if} \quad V_l > K_2 \tag{3.24}$$

where C is the mass flow rate of the refrigerant and K_1 and K_2 represent the volume of the liquid refrigerant inside the vessel. These equations allow the feed to commence when V_l is less than the volume K_1 and continue until the volume of liquid present exceeds K_2.

3.2.9 Liquid separators

Industrial refrigeration systems often have liquid separators and pumps to deliver liquid only to the evaporators at 3 to 5 times the evaporation rate. The return from the evaporators contain both liquid and vapour and is therefore referred to as the wet return. Figure 3.9 shows a typical liquid separator,

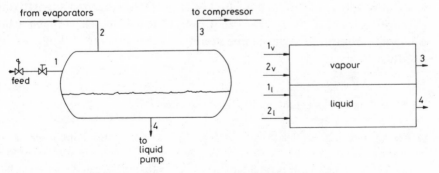

Fig. 3.9 *A liquid separator and the assumed zoning*

which can be represented by two zones, one for the liquid and one for the vapour. The equations that describe the behaviour of the unit and the refrigerant feed are similar to those discussed for the intercooler.

The energy input by the liquid pump can ususally be neglected or the following equation can be used which assumes that the liquid is incompressible.

$$\Delta h = \Delta p / \rho \qquad (3.25)$$

When liquid separators are used it is essential to write equations that allow for the pressure difference between liquid separator and evaporator inlet. The majority of such systems have constant displacement liquid pumps, with the refrigerant delivered to the evaporators at a constant mass flow rate, and this considerably simplifies the calculations.

3.2.10 Final equations for a refrigeration system

With equations having been written to describe the behaviour of each part of the refrigeration system, extensive sorting and manipulation of these equations must follow before a complete mathematical model is available. Some further equations may have to be added to represent the effects of the pipework or if the volumes of these tubes are small and the pressure reduction is not significant, their volumes can be added to those of other components. For example, the tube connecting an intercooler and compressor may be considered as part of the intercooler.

When plant and model responses are compared it is almost always necessary to make adjustments, mainly to heat-transfer coefficients calculated empirically.

When the model is finalized it can be used for many purposes, for example the author has investigated the effect of the quantity of refrigerant charge in the system,[3] obtained performance data for a heat pump,[6] designed control systems,[7] integrated a refrigerator and quick-freeze tunnel model to investigate capacity control[8] and found many other uses for refrigeration system models. Most of this work was carried out at the University of Sheffield under the supervision of Dr. S. A. Marshall, Control Engineering Department, who is now Professor of Mechanical Engineering at the University of Wollongong, Australia.

3.3 Modelling of a vegetable quick-freezing tunnel

The quick-freezing tunnel shown in Fig. 3.10 could be used for freezing a variety of vegetables. The load on the system is fluctuating and depends mainly on the amount of produce being fed into the refrigerated space on the conveyor belt and its temperature. The continuous in-line freezing plant comprises a stainless steel mesh conveyor belt which conveys the produce slowly over the three air-cooling units (evaporators) situated below the return strand of the belt as shown in Fig. 3.10. Seven axial fans drive the air through the evaporators, then through the conveyor belt, and the air is then returned to the fans and recirculated. The quick-freezing tunnel is well insulated. Some air

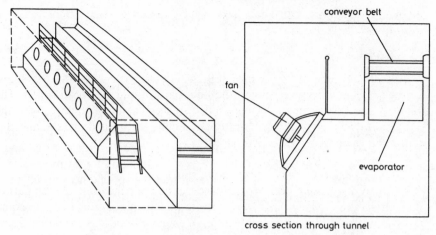

cross section through tunnel

Fig. 3.10 *Layout of the continuous quick-freezing tunnel*

ingress occurs where the conveyor belt passes through the walls and also as a result of various operators opening the door.

The assumed zoning of the quick-freezing tunnel is shown in Fig. 3.11. The air in the refrigerated space was divided into eight zones, these were chosen so that one evaporator fan operated in all but the final zone. The produce on the conveyor belt was similarly divided into eight zones with heat transfer taking place between the produce and air.

The metal in each of the three evaporators was considered as a separate zone with heat transfer taking place between the air and the metal.

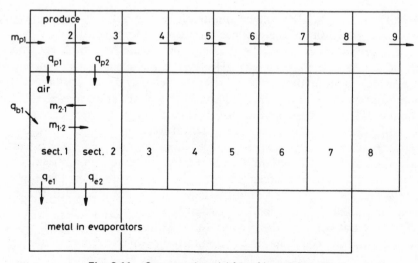

Fig. 3.11 *Conceptual model for refrigerated space*

An energy balance on section 1 (Fig. 1.11), gives

$$q_{p1} + q_{b1} - q_{e1} + m_{2,1}h_2 - m_{1,2}h_1 = \frac{d}{dt}(\rho_{air}h_{air}V_{air}) \qquad (3.26)$$

where q_{p1} is the heat transfer from the produce to the air, q_{b1} is the base load which includes fans, heat leakage etc., q_{e1} is the heat transfer from the air to the metal of the evaporator, $m_{2,1}$ is the mass of air leaving section 2 and entering section 1, and $m_{1,2}$ is the mass of air flowing in the opposite direction. $m_{1,2}$ and $m_{2,1}$ result from turbulence inside the refrigerated space and are assumed to be equal. The volume V_{air} is a constant, and a mean value of $1 \cdot 4 \text{ kg/m}^3$ was used for the density of the air and the specific heat of air was assumed to be unity. Similar equations were then derived for the remaining zones representing the air flow. The mass flow rates $m_{1,2}$ and $m_{2,1}$ were adjusted so that the temperature distribution along the length of the tunnel was the same in both model and plant, and the air flow rates were found to be almost identical between all air zones.

The equations representing the produce were based on surface temperature because it was only possible to measure the mean surface temperature of the produce at the various sections and at entry and exit. An energy and mass balance on the produce passing through section 1 gives

$$K_1(m_{p1}T_{p1} - m_{p2}T_{p2}) - q_{p1} = \rho_1 K_1 \frac{d}{dt}(V_{p1}T_{p1}) \qquad (3.27)$$

and

$$m_{p1} - m_{p2} = \frac{d}{dt}(V_{p1}\rho_{p1}) \qquad (3.28)$$

where V_{p1} and ρ_{p1} are the volume and density of the produce, respectively, in section 1. The density of the produce was assumed to be constant.

If the produce was above the freezing temperature or completely frozen, then K would be the specific heat. However, a phase change is occurring at varying depths from the produce surface depending on its position in the tunnel and on the air temperature. Experience with the plant showed that the produce did not freeze in the first two sections, but water carried into the refrigerated space by the produce would freeze in these sections. It was therefore assumed that an exponential reduction in produce surface temperature occurred through sections 1 to 5 and the constants were evaluated accordingly. For the remaining sections, K was reduced as the freezing approached the core of the produce. Experimental data collected from the plant verifies that this is a good approximation.

The heat transfer from the produce to the air depends on the total surface area, heat-transfer coefficient and the temperature difference, i.e. for section 1

$$q_{p1} = A\alpha(T_{p2} - T_{air}) \qquad (3.29)$$

It was assumed that the surface area varies with the quantity of produce in each section and that the heat-transfer coefficient within a section was constant. Thus $A\alpha$ is proportional to the volume of produce, V_p, and eqn. (3.29) simplifies to:

$$q_{p1} = CV_{p1}(T_{p2} - T_{\text{air}}) \tag{3.30}$$

The constants were evaluated experimentally.

The produce mass flow rate at entry to section 1 was considered to be an input variable, and the mass flow rate at entry to other sections depended on the time taken for the produce to reach the appropriate point, the conveyor belt velocity being constant.

Therefore

$$m_{p1} = \text{input} \tag{3.31}$$

and

$$m_{p2}(t) = m_{p1}(t - c) \tag{3.32}$$

where c is the time taken by the produce passing through section 1. Similar equations were derived for the remaining produce mass flow rates.

The final equations representing the quick-freezing tunnel and the produce passing through it were integrated with a model describing the behaviour of the two-stage ammonia refrigeration system.[8] The resulting mathematical model provided a vast amount of information on the behaviour of the complete process and enabled a capacity control system to be designed, that would conserve energy by ensuring that excess cooling of the produce did not occur as a result of changes in the produce mass flow rate and inlet temperature.

Whilst the equations derived are for a continuous in-line quick-freezing plant, the methods used to derive the model should be applicable to many different types of plants handling a range of produce.

3.4 Modelling of air-conditioning systems

In recent years, mathematical models for a variety of buildings have been developed,[12] however in the absence of suitable air-conditioning-plant models, assumptions have to be made about the internal air temperature that can be maintained. Air-conditioning-plant models have been developed and put to limited use but further work in this field is required.

Hasegawa[9] developed a mathematical dynamical model for a railway passenger car and its-air conditioning system. The model was used to determine the necessary refrigeration capacity, insulation, ventilating and recirculating air volume as well as the transient temperature and humidity inside the passenger car.

Hardy[10] developed a mathematical model of an air-conditioning system for a large lecture theatre. This was an analogue system in which all the energy gains and losses were expressed as an equivalent temperature change in the supply air volume, and the variation of outside air temperature was represented by a sine wave.

The transient responses from the various parts of the plant were assumed to be exponential processes obeying the law

$$\theta_t = \theta_0 \left(1 - e^{-vt}\right) \tag{3.33}$$

where v is the air change rate.

With the increased awareness of the true world energy situation the design of control systems to meet the required conditions within buildings with minimum energy consumption is becoming increasingly important. For this reason mathematical models giving both transient and steady-state responses are likely to play a big role in the design process. The author has therefore chosen to discuss in detail an air-conditioning model which was derived using the stirred-tank approach by James and Marshall.[11] Although this model was for a specific plant the principles involved have a much wider application. Using the same step-by-step technique to derive the model it should be possible for the reader to develop a model for other plants of similar design.

3.4.1 Description of the plant

The air-conditioning plant shown in Fig. 3.12 is used to condition the air in a laboratory in which experiments in human physiology were conducted. Some of the air drawn from the laboratory is exhausted and some passed into a mixing chamber to be mixed with a fresh air supply. The quantity of air recirculated is changed by moving dampers in the inlet, exhaust and recircu-

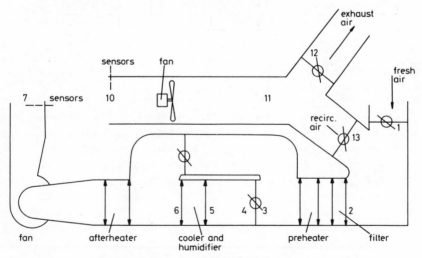

Fig. 3.12   *The air-conditioning plant*

lating air ducts. Servo motors position all three dampers in response to a single hand-operated controller.

The air is passed from the mixing chamber, through a filter and preheater, and is then divided into two streams by servo-positioned dampers, causing a proportion of the air to pass to the humidification and dehumidification chamber and then to a further mixing point to mix with the remaining bypassed air.

The air is then passed through an afterheater into the conditioned space. The refrigeration system employs direct expansion of R12 into the air-cooling unit situated in the duct.

Sensors measure relative humidity and temperature in both supply and return air ducts and are situated between the fans and the conditioned space. Chart recorders measure air temperature at all key positions and Pitot-static tubes measure the air-flow rate. The automatic control system consists of preheater-modulation, humidifier-ON/OFF, refrigerator-ON/OFF, and after-heater-modulation.

The air-conditioning plant is of a much larger capacity than is normally used for a 244 m^3 conditioned space and gives an air change in only 150 seconds.

3.4.2 Development of the mathematical model

The model was first developed by zoning the system of Fig. 3.12. The zones chosen to represent the air-conditioning system are shown in Fig. 3.13. The

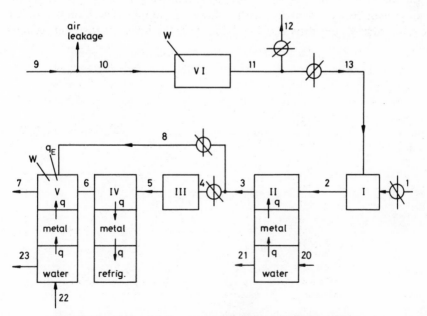

Fig. 3.13 *Zones chosen to represent the air-conditioning system*

equations developed for each zone in turn comprise both algebraic and differential equations and may be categorized as relating to energy, specific humidity, continuity, heat transfer, state and control. The conditioned space was considered to be a single stirred tank. This is obviously an approximation because of the imperfect mixing of the air within the space, however the aim of the study was to investigate the plant behaviour and so the conditioned space was only required to provide a plant load. Hence a large stirred tank with air leakage and heating and cooling was considered adequate to test the model.

Throughout the plant the air was assumed to be incompressible since changes in pressure are small when compared to the total pressure. It follows that in any section, a change in the air mass flow rate at outlet occurs instantaneously if the mass flow rate at inlet is changed. It was also assumed that changes in the water vapour carried by the air does not change the density of mass flow rate of the mixture.

3.4.3 The fresh air mixing chamber
An energy, specific humidity and mass balance on the zone representing the chamber in which the fresh air and recirculated air is mixed (section I, Fig. 3.13) yields the following equations

$$m_1 h_1 + m_{13} h_{13} - m_2 h_2 = V_I \rho_2 \, dh_2/dt \tag{3.34}$$

$$m_1 g_1 + m_{13} g_{13} - m_2 g_2 = V_I \rho_2 \, dg_2/dt \tag{3.35}$$

and

$$m_2 = m_{13} + m_1 \tag{3.36}$$

where g is the specific humidity of the air in kg/kg, ρ is the density of the air/water vapour mixture and other symbols have their previous meaning.

The volume of the fresh-air mixing chamber is constant. Equations giving the density of the air/water vapour mixture, enthalpy, specific humidity and vapour pressure in terms of temperature were established by plotting and fitting and by manipulating the perfect gas laws. The equations so established were

$$g_1 = 0.622 \phi_1 p_{g1}/p_a \tag{3.37}$$

$$h_1 = T_1(1 - g_1) + g_1(2500.0 + 1.9T_1) \tag{3.38}$$

$$\rho_2 = p_a/[R(T_2 + 273.0)] \tag{3.39}$$

$$p_{g1} = 0.0027T_1^2 + 0.0363T_1 + 0.61 \tag{3.40}$$

and

$$g_{g1} = 0.000018T_1^2 + 0.0002T_1 + 0.0083 \tag{3.41}$$

where the suffix g denotes saturation.

3.4.4 Preheater

Three zones were used to represent the preheater (section II) as shown in Fig. 3.13, one for the air between the fresh-air mixing chamber and the dampers which divide the air flow, one for the metal of the externally finned tubes and one for the water inside the tubes. An energy balance on each zone yielded the following equations

air $\qquad m_2 h_2 + q_{2,3} - m_3 h_3 = V_{II} \rho_3 \, dh_3/dt$ $\qquad\qquad$ (3.42)

metal $\qquad q_{20,21} - q_{2,3} = MCp \, dT_M/dt$ $\qquad\qquad\qquad$ (3.43)

water $\qquad m_{20} h_{20} - q_{20,21} - m_{21} h_{21} = V \rho_{21} \, dh_{21}/dt$ $\qquad$ (3.44)

The volume of the air and water zones are constant. An equation giving the density of the air was derived as in the previous section. A mean density for the water inside the tubes was used and assumed to be constant. The specific heat of water was taken to be 4·18 kJ/kg K and that of air unity. The mass of metal was 34·5 kg and the specific heat of copper, 0·39 kJ/kg K.

A specific humidity and mass balance on the zone representing the air yielded the following equations

$$m_2 g_2 - m_3 g_3 = V_{II} \rho_3 \, dg_3/dt \qquad\qquad (3.45)$$

and

$$m_3 = m_2 \qquad\qquad (3.46)$$

It was found necessary to make the heat-transfer coefficient between water and tube a function of the water mass flow rate because this varied considerably, and the following equation giving the heat-transfer rate for water to metal was derived

$$q_{20,21} = 1·74 m_{20}^{0·8}(T_{21} - T_{MP}) \qquad\qquad (3.47)$$

where T_{MP} is the temperature of the metal of the preheater. The outside film coefficient, metal to air, was established experimentally, and the equation so derived was

$$q_{2,3} = 0·91 m_3^{0·8}(T_{MP} - T_3) \qquad\qquad (3.48)$$

3.4.5 The duct connecting the preheater and dehumidifier

An energy, specific humidity and mass balance on this zone (section III, Fig. 3.13), yielded equations similar to (3.42), (3.45) and (3.46). The volume of the zone is constant and equations giving the density and temperature of the air were again established by plotting and fitting and by using the perfect gas laws.

The air is divided into two streams, one entering this zone and one bypassing it. A mass balance on the point where the air stream is divided gives

$$m_4 = m_3 - m_8 \qquad\qquad (3.49)$$

3.4.6 Humidification and dehumidification chamber

Three zones were used to represent this chamber (section IV), one for the air/water vapour mixture between the fins of the cooling coil, one for the metal of the externally finned tubes and one for the refrigerant inside the tubes. During the humidification process the finned tubes are sprayed with water.

An energy, specific humidity and mass balance on the zone representing the air/water vapour mixture yielded the following equations

$$m_5 h_5 - q_{5,6} - m_6 h_6 + m_{wa} h_{wa} - m_{wc} h_{wc} = V_{IV} \rho_6 \, dh_6/dt \tag{3.50}$$

$$m_5 g_5 - m_{wc} + m_{wa} - m_6 g_6 = V_{IV} \rho_6 \, dg_6/dt \tag{3.51}$$

and

$$m_6 = m_5 \tag{3.52}$$

where m_{wc} and m_{wa} are the mass of water vapour condensed or added, respectively.

The volume of the zone is 0.05 m^3, thus indicating that any changes that occur would do so instantaneously and therefore the differential equations (3.50) and (3.51) were replaced by algebraic equations simply by setting the differential coefficients equal to zero.

Since the air adjacent to the metal is cooled more than that farther away, continual mixing occurs, causing the air temperature on leaving the zone to be above the dew point temperature even when dehumidification is taking place. A contact factor, defined as 'the ratio of actual enthalpy drop to the enthalpy drop that would occur if the process on a psychrometric chart continued until the dew point were reached' is therefore used and denoted by β, where

$$\beta = (h_5 - h_6)/(h_5 - h_6') \tag{3.53}$$

The equation can be written in terms of dry bulb temperature with sufficient accuracy for all practical purposes, i.e.

$$\beta = (T_5 - T_6)/(T_5 - T_6') \tag{3.54}$$

An equation giving T_6' was established by plotting and fitting.

During the humidification process the maximum amount of water vapour that the air can retain is given by the equation

$$g_{g5} = 0.00025 h_5 + 0.0005 \tag{3.55}$$

which was established by plotting and fitting. The quantity of water vapour that is added by this unit was established experimentally, giving

$$m_{wa} = 0.65 m_5 (g_{g5} - g_5) \tag{3.56}$$

An energy balance on the metal of the cooling coil gives:

$$q_{5,6} - q_R = M_R Cp \, dT_{MR}/dt \tag{3.57}$$

The mass of metal M_R is 75·6 kg and the specific heat of copper, 0·39 kJ/kg K.

It is known that the refrigerator system responds faster than the air-conditioning system and hence it was assumed that changes taking place in the former did so instantaneously. Algebraic equations were therefore fitted to manufacturers' performance figures giving q_R in terms of the temperature at which the refrigerant is evaporating for a constant condensing temperature of 40°C, and providing a capacity multiplier for other condensing temperatures.

The evaporator heat-transfer coefficient, refrigerant to metal, was established as being 1·14 kW/m² K, giving:

$$q_R = 2 \cdot 4(T_{MR} - T_R) \tag{3.58}$$

The heat-transfer coefficient for metal to air was established experimentally, giving

$$q_{5,6} = 1 \cdot 3 m_6^{0 \cdot 8}(T_5 - T_{MR}) \tag{3.59}$$

For the case when both refrigerator and humidifier were switched off by the control system, the enthalpy, specific humidity and temperature at inlet and outlet of the zone were assumed equal, and for the case when the refrigerator was on but causing no dehumidification, only the specific humidity was assumed equal at inlet and outlet.

3.4.7 Afterheater

The afterheater comprises externally finned tubes heated by water similar to the preheater and six 3-kW auxiliary electric heater elements that are switched on individually.

Three zones were chosen to represent the afterheater section (Fig. 3.13, section V), one for the air, one for the metal in the water heater and one for the water inside the tubes. The air zone included the chamber in which the air bypassing and circulating through the humidification chamber is mixed and the short duct in which the delivery fan is placed. The metal of the electric heaters was not included as a separate zone because their thermal inertia is small and also the number of elements operating is rarely changed while conducting an experiment.

An energy balance on each zone representing the afterheater yielded the following equations:

air $\quad m_6 h_6 + m_8 h_8 + q_{6,7} + q_E + W_{6,7} - m_7 h_7 = V_V \rho_7 \, dh_7/dt$ (3.60)

metal $\quad q_{22,23} - q_{6,7} = M_A Cp \, dt_{MA}/dt$ (3.61)

water $\quad m_{22} h_{22} - q_{22,23} - m_{23} h_{23} = V\rho_{23} \, dh_{23}/dt$ (3.62) where q_E is the output of the auxiliary electric heater elements and W is the fan work.

The volume of the zones containing air and water are constant. Equations giving density, temperature etc., were derived as for the preheater zone. The mass of metal was 18·4 kg and its specific heat 0·39 kJ/kg K. A specific humidity and mass balance on the zone representing the air yielded the following equations

$$m_6 g_6 + m_8 g_8 - m_7 g_7 = V_V \rho_7 \, dg_7/dt \tag{3.63}$$

$$m_7 = m_6 + m_8 \tag{3.64}$$

As with the preheater the heat-transfer coefficient between the water and tube was a function of the water mass flow rate, whilst the heat-transfer coefficient for metal to air was again established experimentally.

3.4.8 Return air duct

An energy, specific humidity, and mass balance on the zone representing the return air duct (section VI, Fig. 3.13), yielded equations similar to (3.60), (3.63) and (3.46), respectively. The volume of the zone is constant and equations giving the density and temperature of the air leaving the zone were derived as previously. The air leaving the zone is divided into two streams, one for the exhaust air and one for the recirculated air. A mass balance on the point where the air stream is divided yields the following equation

$$m_{11} = m_{12} + m_{13} \tag{3.65}$$

3.4.9 Conditioned space

The assumed zoning of the conditioned space is shown in Fig. 3.14. The air inside the conditioned space was considered as a single zone; the walls, floor and ceiling were also considered to be a single zone. This is obviously an approximation because there is imperfect mixing of the air and a temperature variation through the walls.

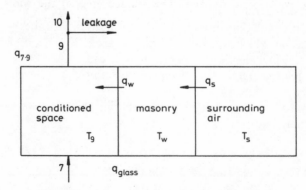

Fig. 3.14 *Assumed zoning of the conditioned space*

An energy, specific humidity, and mass balance on the zone representing the air yielded the following equations:

$$m_7 h_7 + q_{7,9} + q_w + q_{glass} - m_9 h_9 = V\rho_9 \, dh_9/dt \tag{3.66}$$

$$m_7 g_7 - m_9 g_9 = V\rho_9 \, dg_9/dt \tag{3.67}$$

and

$$m_7 = m_9 \tag{3.68}$$

where $q_{7,9}$ is a constant heat load on the room, q_w and q_{glass} are variable heat loads depending on the heat transfer taking place between the room surfaces and the air, and on the heat transfer through the double-glazed windows respectively.

The volume of the air zone is 244 m^3. Equations giving the density and temperature of the air leaving the zone were derived as previously. The relative humidity of the air leaving the zone was obtained from an equation similar to (3.37).

The air leaving the zone is divided into two streams, one entering the return air duct and one to represent air leakage. A mass balance on the point where the air is divided yielded the following equation

$$m_{10} = m_9 - m_1 \tag{3.69}$$

An energy balance on the zone representing the walls, floor and ceiling gave the following equation

$$q_s - q_w = MCp \, dT_W/dt \tag{3.70}$$

The mass of masonry etc., is 66,700 kg and a mean specific heat of 0·65 kJ/kg K was assumed. The suffixes w and s denote masonry and surroundings, respectively.

The heat-transfer coefficient for both sides of the masonry was calculated and modified heat-transfer coefficients which allowed for the thermal conductivity of the walls were derived. The surface area of the walls, floor and ceilings is 227 m^2. The following equations giving the heat-transfer from the masonry to the conditioned space and from the surrounding air to the masonry were thus derived. i.e.

$$q_w = 1\cdot48(T_w - T_9) \tag{3.71}$$

and

$$q_s = 1\cdot48(T_s - T_w) \tag{3.72}$$

The heat transfer between the double-glazed windows and the air inside the conditioned space neglecting the thermal inertia of the glass was given by:

$$q_{glass} = 0\cdot06(T_a - T_9) \tag{3.73}$$

3.4.10 Energy consumption

The energy consumed by the refrigerator was determined experimentally. Over the range of pressures encountered it was found to vary by less than 0·2 kW from the value of 3·5 kW most frequently observed. This value was therefore used in an energy equation and assumed constant when the refrigerator was running. The humidifier spray pump was similarly found to consume 1·5 kW and the fans 0·95 kW. The following energy equation was therefore derived

$$E = \int_0^t (0{\cdot}95 + EH + ER + q_{22,23} + q_{20,21} + q_E)\, dt \qquad (3.74)$$

where EH and ER are the respective instantaneous energy consumptions of the humidifier and refrigerator.

3.4.11 Control system

The hot water mass flow rate through the preheater and afterheater were determined experimentally. The actuating elements were motorized mixing valves. Temperature sensors were situated in the duct immediately after the preheater and in the return air duct providing the required feedback information.

The equations derived were

$$m_{20} = 0{\cdot}12(\text{Const}_P - T_3) \qquad 0 < m_{20} < 0{\cdot}22 \qquad (3.75)$$

and

$$m_{22} = 0{\cdot}06(\text{Const}_A - T_9) \qquad 0 < m_{22} < 0{\cdot}12 \qquad (3.76)$$

Equations were also included to represent valve dynamics.

The humidifier has an ON/OFF controller which also locks off the refrigerator to prevent both working together. Constants C_1 and C_2 are the relative humidity values at which the humidifier pump is switched on and off, respectively.

An ON/OFF controller is used to switch on the refrigerator when required. Constants C_3 and C_4 are the relative humidity upper and lower limits, respectively, and C_5 and C_6 are the upper and lower temperature limits respectively.

The relative humidity ϕ_{10} and the temperature T_{10} which are used for feedback information are both measured by sensors in the return air duct. The sensors can, however, be moved to any position inside the conditioned space. If $\phi_{10} < C_4$ and $T_{10} < C_5$ the refrigerator is off, and if either $\phi_{10} > C_4$ or $T_{10} > C_6$ the refrigerator is on. The flow chart of Fig. 3.15 shows the operation of the control system and how it was programmed.

3.4.12 Final equations

Following extensive sorting and manipulation of the equations, a mathematical model of the air-conditioning system consisting of 18 differential equations

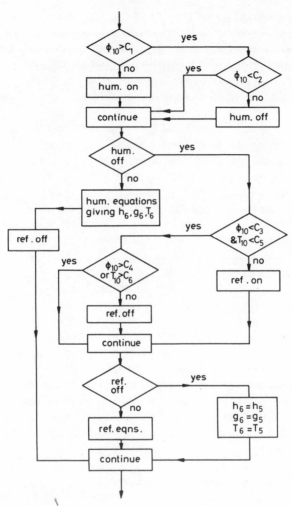

Fig. 3.15 *Flow diagram for control system of humidifier and refrigerator*
Hum.—humidifier
Ref.—refrigerator

and 74 algebraic equations was obtained, and is available from the author. The differential equations were solved digitally using a simple integration routine and at each step all the algebraic equations were solved explicitly. The responses were compared with experimental data. Experience has shown that it is nearly always necessary to modify certain model parameters but in this case because a large amount of experimental data had been obtained, these particular parameters were derived experimentally rather than analytically from the outset. This resulted in the model being a good representation of the plant.

The model provided a vast amount of information on the behaviour of the air-conditioning system with the existing control system during summer and winter operation. The effect of the comfort zone on energy consumption was also investigated. Some of the equations were then changed to simulate the effects of replacing ON/OFF controllers with modulating ones, also the use of specific humidity in place of relative humidity as a control parameter was investigated.

Experience with both model and plant has convinced the author that the parameters which should be controlled are temperature and either vapour pressure or specific humidity. This is not only more in line with the modern thinking among researchers in human physiology, but it makes use of control parameters for the water vapour content of air which are unaffected by temperature, except in the humidification and dehumidification chamber. Two distinct control loops, one for water-vapour content of the air and one for temperature, are then possible with reduced interaction between them. This form of control also reduces total energy consumption.

Serious consideration should also be given to the use of bypass air controllers and enthalpy controllers to operate within the existing control systems. Both are capable of making significant energy savings.

A control system in which the refrigerator is operated by signals from temperature sensors only, with the afterheater remaining off when the temperature of the conditioned space is too high and with the authority of the temperature sensors being transferred to the afterheater only and separate specific humidity sensors then operating the refrigerator when the temperature is reduced, would operate in a more satisfactory manner and conserve energy in the opinion of the author. The model developed is considered ideal for an initial investigation but implementation problems could be encountered.

3.4.13   Modelling of air-conditioned buildings

The environment of any confined space is the result of interaction between the weather conditions, the enclosed structure and heat sources, and sinks within that structure. The importance of designing buildings and plant to work together as environmental modifiers with minimum initial and running costs will eventually lead to dynamic building and plant mathematical models being used together to optimize the design. However, models that simulate the behaviour of buildings are at present large and complex, they are also often used without understanding or realization of the implications of the assumptions that were made during their derivation.

Building services models and building models have been reviewed by Starling.[12] Notable contributions are those of Bradley[13] who investigated the transient behaviour of an experimental masonry building, the Pilkington Glass Company[14] where a dynamic model for buildings has been developed and is being used to simulate office blocks etc., Basnett[15] of the Electricity

Council Research Centre who investigated the behaviour of houses and their heating systems, and Waters[16] who carried out a critical review and developed a very successful model suitable for a variety of buildings.

Buildings are very complex and their thermal behaviour is little understood. If the reader doubts this, he has only to consider how many of the buildings known to him operate properly as climate modifiers and so produce acceptable environmental conditions in all internal areas. Engineers who have an interest in buildings and have access to computer facilities can increase their understanding of the behaviour of various structures by dynamically modelling specific facets of a building and subjecting the models to the various disturbances that can occur as a result of climatic changes. As an example of this technique, a simple mathematical model of a standard double brick wall will now be discussed.

The wall was assumed to consist of standard 100-mm-wide bricks with rendering on the inner surface and was zoned as shown in Fig. 3.16. The render-

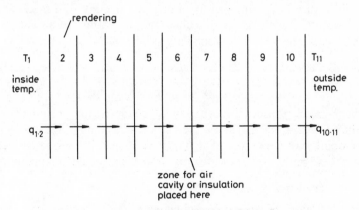

Fig. 3.16 *Assumed zoning of a double brick wall*

ing is seen to be one zone and the double brick wall is represented by eight zones. An energy balance on the rendering gives

$$q_{1,2} - q_{2,3} = MCp \, dT_2/dt \tag{3.77}$$

where $q_{1,2}$ and $q_{2,3}$ are the heat transfer rates per m^2 of wall area between the inside air and rendering, and between the rendering and wall, respectively, and M and Cp are the mass and specific heat of the rendering material. The resistance to heat transfer was assumed to be between the midpoints of neighbouring zones, and for the inner and outer zone it was between the midpoints and the air with the surface heat-transfer coefficients considered. Hence for the rendering

$$q_{1,2} = C_1(T_1 - T_2) \tag{3.78}$$

where T_1 is the inside air temperature, T_2 the zone temperature and C_1 is a constant depending on the zone thickness, thermal conductivity of the material and surface heat-transfer coefficient.

Similar equations were derived for all zones. The energy transmitted through the wall was calculated from the following equation

$$E = \int_0^t q \, dt \qquad (3.79)$$

The model therefore consisted of nine differential equations and eleven algebraic equations with the inside air temperature T_1 and the outside air temperature T_{11} being inputs. With a constant internal temperature of 22°C and the outside air temperature varying sinusoidally between 2 and 14°C over a 24-hour period, the responses from the model for a solid brick wall are shown in Fig. 3.17 and those for a wall with a 50-mm air cavity are shown in Fig. 3.18. These figures show the effect of the air cavity on the variation of wall temperature at varying depths throughout the day. The model was used to investigate the effect of cavity insulation, and 25 mm of insulation on the inner surface of the wall. The results are shown in the table below.

Wall composition	Energy for 24th	Max. heat loss	Min. heat loss
	kJ/m^2	W/m^2	W/m^2
Solid brick wall	2878	41·0	25·0
50-mm air cavity added	1830	26·6	18·0
50-mm insulated cavity	593	8·7	5·2
50-mm air cavity and 25-mm insulation on inner surface	640	8·1	6·3

The energy transmitted per square metre of wall surface area affects the total energy consumption and the maximum heat transmitted affects the heating plant size and therefore capital cost. It is seen from Fig. 3.17 that the maximum heat being absorbed from the room by a solid brick wall occurs at about $6\frac{1}{2}$ hours after the minimum outside temperature is reached. Much can be learned from even a simple model such as this one, for example, the effect on energy consumption of allowing the internal air temperature to fall by a small amount at peak load conditions could be investigated. Similar models for other walls, floors, ceilings etc., are easily derived.

When zoning such systems, choosing zones with equal resistance to heat transfer gives better results than selecting zones of equal thermal capacity.

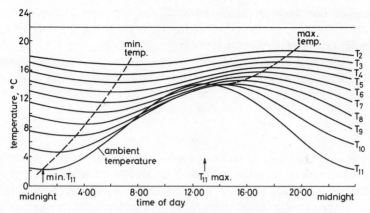

Fig. 3.17 *Variation of wall temperatures with time of day for a solid brick wall*
200-mm solid brick wall
22-mm rendering
No air cavity or insulation
Energy = 2878 kJ/m² for 24 h
Max $q_{1,2} = 41$ Wm² at 07.30
min $q_{1,2} = 25$ Wm² at 19.30

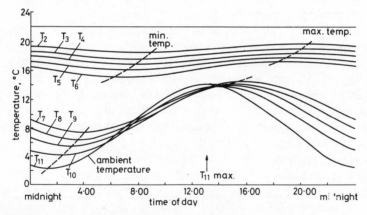

Fig. 3.18 *Variation of wall temperatures with time of day for a cavity wall*
Brick wall of fig. 3.17 with 50-mm air cavity.
No insulation.
Energy = 1830 kJ for 24h
max $q_{1,2} = 26·6$ kWm² at 09.00
min $q_{1,2} = 18$ kWm² at 21.00

However, if the same digital step length is used for all the differential equations, this step length must suit the faster responding equations, although choosing zones on this basis is not economical on computer time. A reasonable compromise between these conflicting requirements can normally be found.

3.5 Unusual uses for mathematical models

Mathematical models for use in the refrigeration and air-conditioning industry are being developed, but compared with other industries, the techniques are still in their infancy. Familiarity with the technique often leads to unusual applications being found for mathematical models. Two examples encountered by the author will now be discussed.

3.5.1 Model of an experimental apparatus

An experimental apparatus to investigate the condensation of pure and mixed refrigerants condensing inside horizontal tubes was designed by Dunn, Saluja and the author.[17] The apparatus was very sophisticated compared with other rigs developed for the same purpose, and consisted of one test and one observation condenser tube, liquid separators, after condensers, liquid subcoolers, a common liquid receiver and pump, four vapour generators and a superheater. The design of the rig removed the compressor characteristics from the refrigerant circuit so that heat transfer, pressure drop and flow regimes could be investigated without the effects of pulsations from the compressor, vibration, oil and other contaminants appearing in the results.

In the long term these parameters will be be replaced in the circuit to ascertain their effects individually and collectively.

The experimental data, such as the various mass flow rates, temperatures and dryness fractions were to be collected with a Datalogger and fed into a computer for statistical analysis. However it was not known how long the apparatus would have to operate before steady-state conditions would be obtained or what would be the best method of controlling pressures, temperature and mass flow rates. Using the stirred-tank approach, a mathematical model consisting of 33 differential equations and 86 algebraic equations with 30 input variables was derived to simulate the rig.

The model was used to estimate the stability of the apparatus over a range of operating conditions and the system response to a change in these conditions. From this data it was possible to determine the minimum interval between taking sets of results and to plan an experimental programme accordingly. The model was also used to confirm that the designed range of variables could be obtained and to indicate the degree of their interaction. Several alternative methods of controlling the system were also investigated.

To derive the model a very detailed study of each part of the apparatus was necessary. This study resulted in a much increased understanding of the fundamental processes of the system which proved invaluable during the later commissioning phase. It also gave the research worker a good insight into the type of information that could be obtained from the rig that would be of use to engineers involved in the mathematical modelling of two-phase fluid-flow systems such as those occurring in refrigeration condensers and evaporators. This proved to be the most valuable part of the exercise.

3.5.2 Model of a banana-boat hold

Bananas are cropped in the Carribean in almost continuous rotation, and supply voyages take place at all times of the year. The bananas are harvested green (unripe) and each stem is inserted into a perforated plastic bag; the average weight of the stems is 50 kg. The stems are loaded directly into the hold and stacked in slatted bins. The ambient temperature is about 35°C but the hold is cooled down before loading commences, the temperature of the fruit must be reduced from 35°C to a carrying temperature of 13°C $\pm$ 1°C with a relative humidity of at least 0·9 within 72 hours of loading. Graphs of respiratory heat variations with fruit temperature are available. The problem was to design the air-conditioning plant for the banana hold and was first seen by the author as a student project.

Calculations showed that 1·5 air changes per hour were required for ventilation of carbon dioxide and of the respiration accelerating gas, ethylene. A large amount of recirculated air would have to pass through the air-conditioning plant because the temperature of the mixed fresh and recirculated air supplied to the hold could not be below 11°C if 90% saturation at 13°C were to be obtained inside the hold. The heat-transfer coefficient between the bins and air was 3·8 W/m² K with a surface area of 1·4 m² for each bin. The mean specific heat was 3·8 kJ/kg K.

An energy balance on the air inside the hold gives the following equation

$$m_1 h_1 + q_1 + q_b - m_1 h_2 = V\rho \, dh_2/dt \qquad (3.80)$$

where m_1 is the mass flow rate of the air entering and leaving the hold, q_1 is the heat leakage through the insulation, q_b is the heat transfer from the slatted bins to the air, and h_1 and h_2 are the enthalpies of the air entering and leaving the hold, respectively.

Similarly an energy balance on the slatted bins gives

$$q_r - q_b = MCp \, dT_b/dt \qquad (3.81)$$

where q_r is the respiratory heat generated inside the bins and M and Cp are the total mass and specific heat of the bins respectively. An equation was fitted to a graph of respiratory heat against produce temperature which give an equation for q_r, and other equations were derived as previously discussed.

The model showed that if the hold was to be maintained at 13°C at all times the air-conditioning plant would be very large. However, by allowing the temperature of the hold to rise initially for a short period then fall again as the produce temperature is reduced, the bananas could still be cooled to 13°C in 72 hours but with a much smaller plant. Steady-state calculations at peak load would have resulted in an oversized plant being chosen and also in increased energy costs.

The use of these simple modelling techniques is of assistance to the systems design engineer but the main problem of designing the plant still remains. However the understanding of the problems has been increased and the engineer will no longer be considering the need to size a plant for peak load conditions to keep the hold at 13°C at all times.

3.6 References

1 BAKER, O.: 'Simultaneous Flow of Oil and Gas', *Oil and Gas J.*, 1954, **53**, p. 185.

2 MARSHALL, S. A.: 'Dynamic Analysis of a Steam Generating Unit', Ph.D. Thesis, University of Cambridge, 1968

3 JAMES, R. W.: 'Modelling and Control of Refrigeration and Air Conditioning Systems for Energy Conservation', Ph.D. Thesis, University of Sheffield, 1976

4 MacLAREN, J. F. T., KERR, S. V., and TRAMSCHEK, A. B.: 'Modelling of Compressors and Valves', *Proc. Inst. Refrig.*, 1974/75, **71**, pp. 42–59

5 DANFOSS AUTOMATIC CONTROLS, DENMARK: 'Liquid Injection into Evaporators', 1967, Pubn. Ref. 100–.8.15.02

6 JAMES, R. W., MARSHALL, S. A., and SALUJA, S. N.: 'The Heat Pump as a means of Utilising Low Grade Heat Energy', *Building Services Engineer*, Jan. 1976, **43**, pp. 202–207

7 MARSHALL, S. A., and JAMES, R. W.: 'An Investigation into the Modelling and Control of an Industrial Refrigeration System', United Kingdom Automation Council, 5th Control Convention, Bath, Sept. 1973

8 MARSHALL, S. A., and JAMES, R. W.: 'Dynamic Analysis of an Industrial Refrigeration System to Investigate Capacity Control', *Proc. Inst. Mech. E.* (*GB*), 1975, **189**(44)

9 HASEGAWA, Y.: 'Analogue Computer Solution of Passenger Car Air Conditioning Process', *SHASE Trans.* (*Japan*), 1965, **2**

10 HARDY, D. J.: 'Simulation on a Digital Computer of an Air Conditioning System Analogue', Technical Memorandum available from the Polytechnic of the South Bank

11 JAMES, R. W., and MARSHALL, S. A.: 'An Investigation into the Control of an Air Conditioning System', Proc. of the Sixth Thermodynamics and Fluid Mechanics Convention, *I. Mech. E.*, Paper No. 15, April 1976

12 STARLING, C.: 'Building Services Computer Programmes Available in the U.K.', Building Research and Information Association publication, 1974

13 BRADLEY, A.: 'Dynamic Thermal Performance of an Experimental Masonry Building', U.S. Dept. of Commerce Publication. Building Sciences Series 45, No. 72-600347, July 1973

14 PILKINGTON GLASS COMPANY: 'Air Conditioning Loads by Computer', Environmental Advisory Service—Pilkington Glass Company, 2nd edition, April 1974

15 BASNETT, P.: 'Mathematical Modelling of the Thermal Behaviour of Houses', 2nd Symposium on the use of Computers for Environmental Engineering related to Buildings, Paris, June 1974, Sect. D, No. D1

16 WATERS, J. R.: 'The Derivation and Experimental Verification of a Computer Aided Thermal Design Method for Buildings', Ph.D. Thesis, Lancaster Polytechnic, 1977

17 DUNN, A., SALUJA, S. N., and JAMES, R. W.: 'An Experimental Rig to Study the Condensation of Pure and Mixed Refrigerants', *Refrigeration and Air Conditioning*, Dec. 1977, pp. 39–47

Chapter 4

Spatial kinetics in nuclear reactor systems

D. H. Owens

4.1 Introduction

The specific problem studied in this chapter is the modelling of oscillations of the power density in a nuclear reactor, such that a 'hot spot' moves from one region of the reactor into another and back again as illustrated schematically in Fig. 4.1. It is assumed that the purpose of the modelling exercise is to investigate the stability of such oscillations both in the open-loop situation and in the presence of feedback or other control action[1-5] regulating the power distribution about a known steady state.

It is not possible in a chapter of this length to give a detailed account of the whole modelling exercise from the basic physical foundations. Rather we take the stance that, as diffusion models are almost universally used by physicists for static and fuel cycling calculations, and that data is hence often available for such models, we should take a diffusion model as a basis for control studies. Such a diffusion model is generally available in the form of a set of coupled, nonlinear partial differential equations in space and time representing the coupled hydraulic, thermodynamic and neutron flux dynamics within the reactor volume. The two obvious difficulties of such models are firstly the nonlinearities and secondly the distributed nature of the system which, if approximated by the use of finite-difference schemes, yields very large numbers of ordinary differential equations if an adequate description of system dynamics is required.

The modelling problem considered in this chapter is the problem of constructing a low-order linear lumped-parameter model of xenon-induced spatial power oscillations in a large, cylindrical nuclear power reactor to replace an (assumed known) nonlinear distributed parameter model. In this context, the idea of 'low-order model' is relative. Typically,[6] a 'low-order model' will consist of several independent sets of 40-90 linear algebraic and ordinary differential equations.

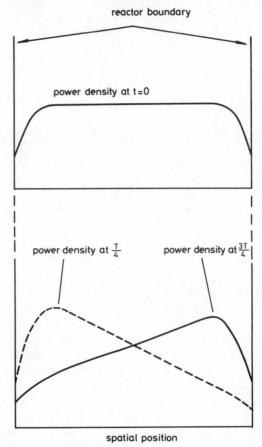

Fig. 4.1   *Power density oscillation of period T in a one-dimensional nuclear reactor*

4.2 Linearization of the diffusion equation

For our purposes a thermal nuclear power reactor will be regarded as a cylindrical volume of space (Fig. 4.2) containing uranium fuel, moderator, coolant, control elements and structural members required to hold the system together. The primary variable of interest is the space-time dependent system power density $p(r, \theta, z, t)$ modelled by relations of the form

$$p(r, \theta, z, t) \triangleq E_0 \int_{E_{\min}}^{E_{\max}} \Sigma_f(r, \theta, z, E)\phi(r, \theta, z, E, t)\, dE$$

$$(\mathrm{MW.cm^{-3}}) \qquad (4.1)$$

where E_0 = average energy produced per fission, $\Sigma_f(r, \theta, z, E)$ is the macroscopic fission cross section at the position (r, θ, z) for neutrons of energy E,

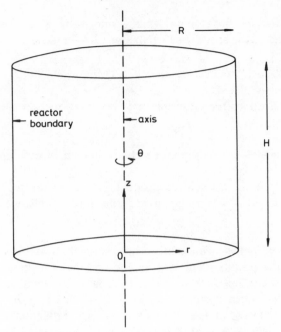

Fig. 4.2   *Reactor volume and spatial coordinates*

and the neutron flux at the point (r, θ, z) due to neutrons in any energy range $E_1 \leq E \leq E_2$ is represented by

$$\phi(r, \theta, z, E_1, E_2, t) \triangleq \int_{E_1}^{E_2} \phi(r, \theta, z, E, t)\, dE \qquad (\text{n.cm}^{-2}.\text{s}^{-1}) \qquad (4.2)$$

In practice, it is not possible to obtain data on the fission cross sections over the whole energy range. The power distribution is modelled by supposing that the energy continuum can be approximated by a number of discrete energy bands $E_{min} = E_1 \leq E \leq E_2$, $E_2 \leq E \leq E_3$, ..., $E_l \leq E \leq E_{l+1} = E_{max}$ and averaging over these energy bands. For example, using the notation

$$\phi_j(r, \theta, z, t) \triangleq \phi(r, \theta, z, E_j, E_{j+1}, t) \qquad 1 \leq j \leq l \qquad (4.3)$$

then eqn. (4.1) is approximated by the linear form

$$p(r, z, \theta, t) = E_0 \sum_{j=1}^{l} \Sigma_j(r, \theta, z)\phi_j(r, \theta, z, t) \qquad (4.4)$$

where the $\Sigma_j(r, \theta, z)$ are average fission cross sections in the range $E_j \leq E \leq E_{j+1}$ and are estimated experimentally. The space-time behaviour of the ϕ_j are then modelled by the coupled set of 'multigroup' diffusion equations.[1-5]

For illustrative purposes consider the one-group case $(l = 1)$ and the simplified model of the space-time behaviour of $\phi_1(r, \theta, z, t)$

$$\nabla^2 \phi(r, \theta, z, t) + B^2(r, \theta, z, t)\phi(r, \theta, z, t) = \frac{l^*}{M^2} \frac{\partial \phi}{\partial t}(r, \theta, z, t) \qquad (4.5)$$

where we have dropped the subscript on ϕ_1 for notational simplicity. Term by term

(a) the Laplacian term $\nabla^2 \phi$ represents diffusion of neutrons throughout the reactor volume
(b) the term $B^2 \phi$ is the net neutron production rate per unit volume, and
(c) the term $(l^*/M^2) \partial \phi/\partial t$ represents the rate of change of neutron population.

The coefficient l^* is the mean neutron lifetime in the reactor core. The parameter M^2 is the position-dependent migration area of neutrons in the reactor core. The coefficient $B^2(r, \theta, z, t)$ is the space-time (and, as we shall see later, power) dependent geometric buckling of the flux at (r, θ, z) and time t. In principle therefore, eqn. (4.5) is a nonlinear partial differential equation.

The system is also subjected to boundary conditions of the form

$$\phi(R, \theta, z, t) + l_R \frac{\partial \phi}{\partial r}(r, \theta, z, t)\bigg|_{r=R} \equiv 0$$

$$\phi(r, \theta, 0, t) - l_z \frac{\partial \phi}{\partial z}(r, \theta, z, t)\bigg|_{z=0} \equiv 0$$

$$\phi(r, \theta, H, t) + l_z \frac{\partial \phi}{\partial z}(r, \theta, z, t)\bigg|_{z=H} \equiv 0 \qquad (4.6)$$

where l_R and l_z are the radial and axial extrapolation distances, respectively, and the rotational invariance condition

$$\phi(r, \theta, z, t) \equiv \phi(r, \theta + 2\pi, z, t) \qquad (4.7)$$

The boundary conditions (4.6) are only an approximation to the real physical situation and are based on the construction illustrated in Fig. 4.3 for the radial case, i.e. it is assumed that the neutron flux vanishes at a fixed point $R + l_R$ and that this point coincides with that obtained by linear extrapolation of the neutron flux from the reactor boundary.

As the objective is to obtain a linear model of the dynamics of small spatial power perturbations about a fixed operating point, the first job is to characterize the reactor steady state. More precisely, if $B_0^2(r, \theta, z)$ and $\phi_0(r, \theta, z)$ are the steady-state buckling and flux distributions, respectively, eqn. (4.5) reduces to the search for *positive* solutions of the partial differential equation

$$\nabla^2 \phi_0(r, \theta, z) + B_0^2(r, \theta, z)\phi_0(r, \theta, z) \equiv 0 \qquad (4.8)$$

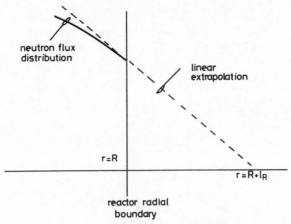

Fig. 4.3 *Boundary condition construction*

with boundary conditions easily derived from eqn. (4.6). Note, in particular, that the steady state is characterized and governed by the geometric buckling and the boundary conditions.

Following normal procedures, write all variables as the sum of steady-state values plus a perturbation from the steady state

$$\phi(r, \theta, z, t) = \phi_0(r, \theta, z) + \delta\phi(r, \theta, z, t)$$
$$B^2(r, \theta, z, t) = B_0^2(r, \theta, z) + \delta B^2(r, \theta, z, t) \qquad (4.9)$$

Substituting into eqn. (4.5), using the steady-state condition (4.8) and neglecting the product term $\delta B^2 \, \delta\phi$ yields the following approximate linear model for small transient perturbations about the specified reactor steady state

$$\nabla^2 \, \delta\phi(r, \theta, z, t) + B_0^2(r, \theta, z) \, \delta\phi(r, \theta, z, t)$$
$$+ \phi_0(r, \theta, z) \, \delta B^2(r, \theta, z, t) = \frac{l^*}{M^2} \frac{\partial \delta\phi}{\partial t} (r, \theta, z, t) \qquad (4.10)$$

The reader should easily verify that $\delta\phi$ is subject to the boundary conditions of the form of eqn. (4.6) with ϕ replaced by the perturbation $\delta\phi$.

Finally, if we restrict our attention to slow, long-term transients the term $(l^*/M^2) \, \partial\delta\phi/\partial t$ can be neglected leaving the model

$$\nabla^2 \, \delta\phi(r, \theta, z, t) + B_0^2(r, \theta, z) \, \delta\phi(r, \theta, z, t)$$
$$+ \phi_0(r, \theta, z) \, \delta B^2(r, \theta, z, t) = 0 \qquad (4.11)$$

4.3 Linearization of the iodine/xenon equations

A major destabilizing influence[4-6] on large thermal reactor systems is the slow dynamic effect of the fission xenon-135 produced by the decay chain

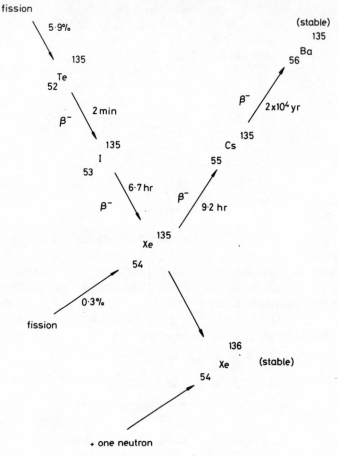

Fig. 4.4　*Xenon decay chain*

shown in Fig. 4.4. The significance of xenon lies in the dual property of being produced by neutron bombardment of uranium-235 at a rate proportional to the neutron flux and its simultaneous decay by β-decay and destruction by absorption of neutrons at a rate proportional to the product of the neutron flux and the xenon concentration (cm^{-3}). In effect the iodine/xenon dynamics are a nonlinear inherent feedback loop within the reactor system.

Restricting attention to slow, long-term transients (with periods measured in hours) the Te-135 transition can be neglected in the xenon decay chain. The decay chain can then be represented by the two partial differential equations

$$\gamma_I \Sigma_f(r, \theta, z)\phi(r, \theta, z, t) - \lambda_I I(r, \theta, z, t) = \frac{\partial I}{\partial t}(r, \theta, z, t) \qquad (4.12)$$

and

$$\gamma_\chi \Sigma_f(r, \theta, z)\phi(r, \theta, z, t) + \lambda_I I(r, \theta, z, t)$$

$$- (\lambda_\chi + \sigma_\chi \phi(r, \theta, z, t))X(r, \theta, z, t) = \frac{\partial X}{\partial t}(r, \theta, z, t) \qquad (4.13)$$

where γ_I and γ_χ are the fractional yields of iodine and xenon in fission respectively, λ_I and λ_χ are the iodine and xenon nuclear decay constants respectively, σ_χ is the microscopic absorption cross section of xenon-135 for thermal neutrons and $I(r, \theta, z, t)$ and $X(r, \theta, z, t)$ are the space-time concentrations (cm^{-3}) of iodine and xenon respectively in the reactor core.

It is easily verified that the steady-state iodine and xenon concentrations corresponding to the steady-state neutron flux distribution ϕ_0 are given by the formulae

$$I_0(r, \theta, z) = \gamma_I \Sigma_f(r, \theta, z)\phi_0(r, \theta, z)/\lambda_I$$

$$X_0(r, \theta, z) = \frac{(\gamma_I + \gamma_\chi)\Sigma_f(r, \theta, z)\phi_0(r, \theta, z)}{\lambda_\chi + \sigma_\chi \phi_0(r, \theta, z)} \qquad (4.14)$$

respectively. Writing

$$I(r, \theta, z, t) = I_0(r, \theta, z) + \delta I(r, \theta, z, t)$$

$$X(r, \theta, z, t) = X_0(r, \theta, z) + \delta X(r, \theta, z, t) \qquad (4.15)$$

then the linearized versions of eqns. (4.12) and (4.13) take the form

$$\gamma_I \Sigma_f(r, \theta, z)\,\delta\phi(r, \theta, z, t) - \lambda_I\,\delta I(r, \theta, z, t) = \frac{\partial\delta I}{\partial t}(r, \theta, z, t) \qquad (4.16)$$

and

$$\{\gamma_\chi \Sigma_f(r, \theta, z) - \sigma_\chi X_0(r, \theta, z)\}\,\delta\phi(r, \theta, z, t) + \lambda_I\,\delta I(r, \theta, z, t)$$

$$- \{\lambda_\chi + \sigma_\chi \phi_0(r, \theta, z)\}\,\delta X(r, \theta, z, t) = \frac{\partial\delta X}{\partial t}(r, \theta, z, t) \qquad (4.17)$$

As these equations contain no spatial derivatives, there are no spatial boundary conditions on δI and δX.

4.4 Inherent feedback and the homogeneous model

The final step in the construction of the linear, distributed parameter model of slow, small transient spatial power perturbations about the specified steady state is the characterization of the buckling perturbation δB^2 occurring in eqn. (4.11). This is a highly complex task as this change is a complex non-linear function of reactor temperature distributions, coolant dynamics, xenon concentrations and control action. A common approach is to argue that all dynamic effects other than xenon will have reached steady state if we consider slow transients only, suggesting that their effects can be represented by a simple gain, i.e. our model is

$$\delta B^2(r, \theta, z, t) \triangleq K_p \, \delta\phi(r, \theta, z, t) + K_X \, \delta X(r, \theta, z, t) + u(r, \theta, z, t) \quad (4.18)$$

where $u(r, \theta, z, t)$ is a space-time independent control input function representing the effect of control rods or other control devices, K_X is the xenon 'reactivity coefficient' and K_p is the reactor 'power coefficient'. Both K_X and K_p could be position dependent but, for our purposes, it is assumed that they are constant.

The final homogeneous model (obtained by setting the control action $u \equiv 0$) takes the form of eqns. (4.11), (4.16), (4.17) and (4.18) or the operator-theoretic form

$$L\phi = \mu \frac{\partial \phi}{\partial t} \quad (4.19)$$

where the space-dependent operator

$$L \triangleq \begin{pmatrix} \nabla^2 + B_0^2 + \phi_0 K_p & 0 & \phi_0 K_X \\ \gamma_I \Sigma_f & -\lambda_I & 0 \\ \gamma_\chi \Sigma_f - \sigma_\chi X_0 & \lambda_I & -(\lambda_\chi + \sigma_\chi \phi_0) \end{pmatrix} \quad (4.20)$$

and

$$\mu \triangleq \begin{pmatrix} 0 & 0 & 0 \\ 0 & 1 & 0 \\ 0 & 0 & 1 \end{pmatrix}$$

$$\phi^T \triangleq (\delta\phi(r, \theta, z, t), \delta I(r, \theta, z, t), \delta X(r, \theta, z, t))^T \quad (4.21)$$

together with spatial boundary conditions on $\delta\phi$. The feedback structure of the system is illustrated in Fig. 4.5 and indicates the potentially destabilizing effects of both the xenon and power feedback loops.

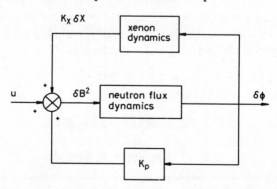

Fig. 4.5 *Inherent feedback loops in long-term reactor spatial dynamics*

4.5 Xenon-induced instability: analytical methods

Despite its formal simplicity, the linear model is not in a form suitable for stability calculations. It can, however, be used to suggest useful approximation schemes and general procedures. The investigation of stability is initiated by looking for nontrivial exponential solutions of eqn. (4.19) of the form

$$\phi(r, \theta, z, t) = \phi_\lambda(r, \theta, z)e^{\lambda t} \qquad (4.22)$$

and assuming that the linear model is asymptotically stable if and only if all solutions satisfy Re $\lambda < 0$. Substituting eqn. (4.22) into (4.19) yields

$$L\phi_\lambda = \lambda\mu\phi_\lambda \qquad (4.23)$$

or, equivalently, the stability problem can be regarded as the evaluation of the dominant generalized eigenvalues of L. This must be undertaken numerically.

The numerical problems arising in the analysis of eqn. (4.23) can be very severe unless care is given in the choice of numerical approximations. The use of finite difference approximations to the spatial derivatives occurring in L can be immediately eliminated owing to excessive dimensionality. For example, reasonable accuracy would require of the order of 150 mesh points in the (r, θ) plane and 10 in the axial direction which, bearing in mind the fact that there are three variables at each mesh point, leads to a generalized eigenvalue problem of dimension $150 \times 10 \times 3 = 4500$. In this section we consider how modal analysis methods can help reduce this severe problem.

It is clear that knowledge of ϕ_λ is equivalent to knowledge of λ. It is also clear from the complex space-dependence of the elements of L that the analytical determination of ϕ_λ is an impossible task. We therefore attempt to find a solution by expanding ϕ_λ as the sum of known functions.[5, 7, 8]

$$\phi_\lambda = \sum_{j=1}^{\infty} v_j \psi_j(r, \theta, z) \qquad (v_j \in R^3, j \geq 1) \qquad (4.24)$$

where, for example, the scalar functions ψ_j are the eigenfunctions of the operator $\nabla^2 + B_0^2$, i.e.

$$\nabla^2 \psi_j + B_0^2 \psi_j = \gamma_j \psi_j \qquad j \geq 1 \qquad (4.25)$$

In particular we can guarantee that ϕ_λ satisfies the spatial boundary conditions if we ensure that each ψ_j satisfies the boundary conditions.

Taking for simplicity the case of $l_R = l_z = 0$, the identity

$$(\gamma_j - \gamma_k) \int_V \psi_j \psi_k \, dV = \int_V \{\psi_j \nabla^2 \psi_k - \psi_k \nabla^2 \psi_j\} \, dV$$

$$= \int_V \text{div} \{\psi_j \nabla\psi_k - \psi_k \nabla\psi_j\} \, dV$$

$$= \int_{\substack{\text{reactor} \\ \text{boundary}}} \{\psi_j \nabla\psi_k - \psi_k \nabla\psi_j\}.ds$$

$$= 0 \qquad (4.26)$$

indicates that we can assume that the set $\{\psi_j\}_{j1}$ is orthonormal, i.e.

$$\int_V \psi_j \psi_k \, dV = \delta_{jk} \tag{4.27}$$

where δ_{jk} is the Kronecker delta and V is the volume of the reactor core. The following development will, of course, be valid for any orthonormal choice of the $\{\psi_j\}_{j \geq 1}$. In particular we note that any choice of linearly independent ψ_j can be replaced by an orthonormal set using Gram-Schmidt orthonormalization.[11]

Substituting eqn. (4.24) into (4.23) yields the relation

$$\sum_{j=1}^{\infty} L\psi_j v_j = \lambda\mu \sum_{j=1}^{\infty} \psi_j v_j \tag{4.28}$$

multiplying by ψ_k, integrating over the reactor volume and using eqn. (4.27) yields the relations

$$\sum_{j=1}^{\infty} L_{kj} v_j = \lambda\mu v_k \qquad k \geq 1 \tag{4.29}$$

where the 3×3 constant matrices

$$L_{kj} \triangleq \int_V \psi_k L\psi_j \, dV$$

$$= \int_V \psi_k \begin{pmatrix} \gamma_j + \phi_0 K_p & 0 & \phi_0 K_X \\ \gamma_I \Sigma_f & -\lambda_I & 0 \\ \gamma_\chi \Sigma_f - \sigma_\chi X_0 & \lambda_I & -(\lambda_\chi + \sigma_\chi \phi_0) \end{pmatrix} \psi_j \, dV \tag{4.30}$$

are easily evaluated numerically.

The infinite set of equations are still numerically intractable but approximate solutions can be obtained by a variety of means.

Method 1: If the elements of the 3×3 matrix appearing in the integral of eqn. (4.30) are constant over a large volume of the reactor it is intuitively plausible that

$$L_{jk} \simeq 0 \qquad j \neq k \tag{4.31}$$

when the eigenvalue eqn. (4.29) has the approximate, but highly simple, form

$$L_{kk} v_k = \lambda\mu v_k \qquad k \geq 1 \tag{4.32}$$

Equivalently the eigenvalues are the solutions of relations of the form

$$|L_{kk} - \lambda\mu| = 0 \tag{4.33}$$

which are quadratics in λ that are easily analysed by pencil and paper methods. The approximation of eqn. (4.31) is fairly severe however, and the

results can only be regarded as giving rough estimates and indicating parametric trends.

Method 2: If the series in the above converge rapidly enough, eqn. (4.29)
can be approximated by a truncated form

$$\sum_{j=1}^{M} L_{kj} v_j = \lambda \mu v_k \qquad 1 \le k \le M \tag{4.34}$$

or, equivalently, the $3M \times 3M$ generalized eigenvalue problem

$$\begin{pmatrix} L_{11} & L_{12} & \cdots & L_{1M} \\ \vdots & & & \vdots \\ L_{M1} & & \cdots & L_{MM} \end{pmatrix} \begin{pmatrix} v_1 \\ \vdots \\ v_M \end{pmatrix} = \lambda \begin{pmatrix} \mu & 0 & \cdots & 0 \\ 0 & \mu & & \\ \vdots & & & 0 \\ 0 & \cdots & 0 & \mu \end{pmatrix} \begin{pmatrix} v_1 \\ \vdots \\ v_M \end{pmatrix} \tag{4.35}$$

for which known solution methods exist.[9] For example, it is observed that
eqn. (4.35) is equivalent to the determinantal relation

$$\det \left\{ \begin{pmatrix} L_{11} & \cdots & L_{1M} \\ \vdots & & \vdots \\ L_{M1} & \cdots & L_{MM} \end{pmatrix} - \lambda \begin{pmatrix} \mu & \cdots & 0 \\ \vdots & & \vdots \\ 0 & \cdots & \mu \end{pmatrix} \right\} = 0 \tag{4.36}$$

which, by suitable row and column operations, reduces to the form

$$\det \left\{ \begin{pmatrix} A_{11} & A_{12} \\ A_{21} & A_{22} - \lambda I_{2M} \end{pmatrix} \right\} = 0 \tag{4.37}$$

or, if $|A_{11}| \ne 0$,

$$|\lambda I_{2M} - A_{22} + A_{21} A_{11}^{-1} A_{12}| = 0 \tag{4.38}$$

which is a $2M \times 2M$ standard eigenvalue problem. In practice good estimates
of the stability of the configuration can be achieved using fairly small numbers of ψ_j, typically 20–40. The corresponding eigenvalue calculations are
hence of dimension 40–80 which is fairly large, but manageable as, in
practice,[6, 10] the matrices are fairly well conditioned.

4.6 Xenon-induced transients: state space models

The techniques described in the previous section for obtaining models for use
in stability calculations based on eigenfunction (or 'modal') methods is easily
extended[5, 8] to produce state space models[12] describing the reactor response
to control inputs. In general terms, suppose that the control input distribution
$u(r, \theta, z, t)$ can be represented as the sum of m independent contributions with
a given spatial distribution but manipulable amplitudes, i.e.

$$u(r, \theta, z, t) = \sum_{j=1}^{m} f_j(r, \theta, z) u_j(t) \tag{4.39}$$

and let $u(t) = (u_1(t), \ldots, u_m(t))^T$ be the system manipulable input vector. The input-driven model can now be obtained from eqns. (4.11) and (4.16) to (4.18) to be of the form

$$\mu \frac{\partial \phi}{\partial t} = L\phi + Fu \qquad (4.40)$$

where

$$F = \begin{pmatrix} \phi_0 f_1 & \phi_0 f_2 & \cdots & \phi_0 f_m \\ 0 & 0 & \cdots & 0 \\ 0 & 0 & \cdots & 0 \end{pmatrix} \qquad (4.41)$$

In a similar manner to eqn. (4.24) suppose that ϕ can be expanded as a linear combination of known orthonormal functions $\psi_j(r, \theta, z)$, $j \geq 1$, with time dependent amplitudes $v_j(t) \in R^3$, i.e.

$$\phi = \sum_{j=1}^{\infty} \psi_j(r, \theta, z)v_j(t) \qquad (4.42)$$

Substituting into eqn. (4.40), similar techniques to those used in Section 4.5 can be used to replace this model by an infinite set of first-order vector ordinary differential equations

$$\mu \frac{dv_k(t)}{dt} = \sum_{j=1}^{\infty} L_{kj}v_j(t) + F_k u(t) \qquad k \geq 1 \qquad (4.43)$$

where F_k is the $3 \times m$ constant matrix

$$F_k \triangleq \int_V \psi_k F \, dV \qquad k \geq 1 \qquad (4.44)$$

A finite-dimensional state space model approximating (4.43) is obtained by the truncation technique of method II of Section 4.5, i.e. replace (4.43) by the truncated set

$$\mu \frac{dv_k(t)}{dt} = \sum_{j=1}^{M} L_{kj}v_j(t) + F_k u(t) \qquad 1 \leq k \leq M \qquad (4.45)$$

which can be written in the (nonstandard) form

$$\mu_e \dot{x}(t) = Ax(t) + Bu(t) \qquad (4.46)$$

where

$$x(t) = (v_1^T(t), v_2^T(t), \ldots, v_M^T(t))^T$$

$$A = \begin{pmatrix} L_{11} & \cdots & L_{1M} \\ \vdots & & \vdots \\ L_{M1} & \cdots & L_{MM} \end{pmatrix} \qquad B = \begin{pmatrix} F_1 \\ \vdots \\ F_M \end{pmatrix} \qquad \mu_e = \begin{pmatrix} \mu & 0 & \cdots & 0 \\ 0 & & & \vdots \\ \vdots & & & 0 \\ 0 & & 0 & \mu \end{pmatrix}$$

$$(4.47)$$

If we also suppose that m output measurements of the form, $1 \leq k \leq m$,

$$y_k(t) = \int_{V_k} p(r, \theta, z, t) \, dV \tag{4.48}$$

equal to the total power generated in a colume V_k of the reactor core, then substituting from eqns. (4.4) and (4.42) we obtain

$$y_k(t) = \sum_{j=1}^{\infty} C_{kj} v_j(t)$$

$$C_{kj} = E_0 \int_{V_k} \Sigma_f(r, \theta, z) \psi_j(r, \theta, z) \, dV \quad (1 \ 0 \ 0) \tag{4.49}$$

Truncating after M terms, and defining the output vector $y(t) = (y_1(t), \ldots, y_m(t))^T$ we obtain the standard form of output equation

$$y(t) = Cx(t) \tag{4.50}$$

where

$$C = \begin{pmatrix} C_{11} & \cdots & C_{1M} \\ \vdots & & \vdots \\ C_{m1} & \cdots & C_{mM} \end{pmatrix} \tag{4.51}$$

The combined state space model of eqns. (4.46) and (4.50) can be used for simulation purposes, the accuracy increasing as M increases. The model can also be used as the basis of control studies[4–6, 8, 12] based on time-domain control synthesis procedures. Controller design could also proceed[6] based on multivariable frequency domain design techniques[12] using the reactor transfer function matrix

$$G(s) = C(s\mu_e - A)^{-1} B \tag{4.52}$$

(obtained by taking Laplace transforms of eqns. (4.46) and (4.50) with zero initial conditions).

4.7 Important case of azimuthal symmetry

The use of eigenfunction/modal methods makes possible a significant reduction in model dimension when compared with finite-difference models but, for good results, requires careful choice of functions ψ_j, $1 \leq j \leq M$. The use of the eigenfunction defined by eqn. (4.25) is a good choice but they must, in general, be computed numerically. This is a difficult task in its own right. In practice, therefore, a compromise can be reached by representing certain spatial distributions by modal expansions (when such modal expansions can be easily defined) and other spatial distributions by finite-difference approximation schemes.[10] A case of particular interest is described below.

In many practical situations, the reactor steady state is azimuthally symmetric in the sense that the steady-state flux and fuel distributions $\phi_0(r, \theta, z)$ and $\Sigma_f(r, \theta, z)$ are independent of θ. This immediately has some implications for the structure of the solutions of eqn. (4.40). More precisely, remembering that the solutions are periodic in θ [eqn. (5.7)], expand ϕ as a Fourier series

$$\phi(r, \theta, z, t) = \psi_0(r, z, t) + \sum_{k=1}^{\infty} \{\psi_{jc}(r, z, t) \cos n\theta + \psi_{js}(r, z, t) \sin n\theta\}$$

(4.53)

where $\psi_0(r, z, t)$, $\psi_{jc}(r, z, t)$ and $\psi_{js}(r, z, t)$ are unknown functions to be determined. Substituting into eqn. (4.40) and noting that

$$\frac{\partial^2}{\partial \theta^2}\begin{pmatrix} \cos n\theta \\ \sin n\theta \end{pmatrix} = -n^2 \begin{pmatrix} \cos n\theta \\ \sin n\theta \end{pmatrix} \qquad n \geq 0$$

(4.54)

yields the relations (after a little manipulation)

$$\mu \frac{\partial \psi_0}{\partial t} + \sum_{k=1}^{\infty} \left\{ \mu \frac{\partial \psi_{jc}}{\partial t} \cos n\theta + \mu \frac{\partial \psi_{js}}{\partial t} \sin n\theta \right\}$$

$$= L_0 \psi_0 + \sum_{k=1}^{\infty} \{\cos n\theta L_n \psi_{jc} + \sin n\theta L_n \psi_{js}\} + Fu$$

(4.55)

where, for $n \geq 0$, the operator L_n is obtained from L by replacing ∇^2 by

$$\frac{1}{r} \frac{\partial}{\partial r} r \frac{\partial}{\partial r} + \frac{\partial^2}{\partial z^2} - \frac{n^2}{r^2}$$

(4.56)

Defining the Fourier coefficients of F

$$F_0 = \frac{1}{2\pi} \int_0^{2\pi} F(r, \theta, z) \, d\theta$$

$$F_{nc} = \frac{1}{\pi} \int_0^{2\pi} F(r, \theta, z) \cos n\theta \, d\theta \qquad n \geq 1$$

(4.57)

$$F_{ns} = \frac{1}{\pi} \int_0^{2\pi} F(r, \theta, z) \sin n\theta \, d\theta \qquad n \geq 1$$

then the linear independence of the trigonometric functions indicates that the model of eqn. (4.40) can be reduced to the model

$$\mu \frac{\partial \psi_0}{\partial t} = L_0 \psi_0 + F_0 u$$

$$\mu \frac{\partial \psi_{nc}}{\partial t} = L_n \psi_{nc} + F_{nc} u \qquad n \geq 1$$

(4.58)

$$\mu \frac{\partial \psi_{ns}}{\partial t} = L_n \psi_{ns} + F_{ns} u \qquad n \geq 1$$

Assuming, for simplicity, zero extrapolation lengths, the spatial boundary conditions on these equations take the form

$$\psi_0(R, z, t) \equiv \psi_{nc}(R, z, t) \equiv \psi_{ns}(R, z, t) \equiv 0 \qquad n \geq 1$$

$$\psi_0(r, 0, t) \equiv \psi_{nc}(r, 0, t) \equiv \psi_{ns}(r, 0, t) \equiv 0 \qquad n \geq 1 \qquad (4.59)$$

$$\psi_0(r, H, t) \equiv \psi_{nc}(r, H, t) \equiv \psi_{ns}(r, H, t) \equiv 0 \qquad n \geq 1$$

together with continuity and differentiability requirements on the axis ($r = 0$),

$$\psi_{nc}(0, z, t) \equiv \psi_{ns}(0, z, t) \equiv 0 \qquad n \geq 1$$

$$\left. \frac{\partial \psi_{nc}}{\partial r}(r, z, t) \right|_{r=0} \equiv \left. \frac{\partial \psi_{ns}}{\partial r}(r, z, t) \right|_{r=0} \equiv 0 \qquad n = 2k \qquad k \geq 0 \qquad (4.60)$$

The output eqn. (4.48) takes the vector form

$$y_k(t) = y_0(t) + \sum_{n=1}^{\infty} \{y_{nc}(t) + y_{ns}(t)\} \qquad (4.61)$$

where

$$(y_0(t))_k \triangleq E_0 \int_{V_k} \Sigma_f(r, z) \psi_0(r, z, t) \, dV$$

$$(y_{nc}(t))_k \triangleq E_0 \int_{V_k} \Sigma_f(r, z) \cos n\theta \psi_{nc}(r, z, t) \, dV \qquad n \geq 1 \qquad (4.62)$$

$$(y_{ns}(t))_k \triangleq E_0 \int_{V_k} \Sigma_f(r, z) \sin n\theta \psi_{ns}(r, z, t) \, dV \qquad n \geq 1$$

represent the contributions to the kth output from the various trigonometric modes.

It is important to note that the decomposition of the model expressed by eqns. (4.58) and (4.61) represents a significant potential numerical advantage. For example, in stability studies, it is easily seen that solutions λ of the generalized eigenvalue problem eqn. (4.23) are also solutions of generalized eigenvalue problems of the form of

$$L_n \eta_n(r, z) = \lambda \mu \eta_n(r, z) \qquad (4.63)$$

for some $n \geq 0$. Conversely, if λ satisfies eqn. (4.63), it also satisfies eqn. (4.23) with $\phi_\lambda = \eta_n \cos n\theta$ or $\phi_\lambda = \eta_n \sin n\theta$. Moreover, it can be shown[6] that, in rough terms,

$$\lim_{n \to +\infty} \lambda < 0 \qquad (4.64)$$

and hence, for stability studies, the eigenvalue eqn. (4.63) need only be considered in some range $0 \leq n \leq M$ (typically 2 or 3).

In a similar manner, it can be shown[6] that the series in eqn. (4.53) converges at least as fast as $0\,(1/n^2)$ and hence, without too much loss of accuracy, can be truncated to yield the approximate model

$$\mu\frac{\partial\psi_0}{\partial t} = L_0\psi_0 + F_0 u$$

$$\mu\frac{\partial\psi_{nc}}{\partial t} = L_n\psi_{nc} + F_{nc}u \qquad 1 \le n \le M \tag{4.65}$$

$$\mu\frac{\partial\psi_{ns}}{\partial t} = L_n\psi_{ns} + F_{ns}u \qquad 1 \le n \le M$$

with the approximate output model

$$y(t) = y_0(t) + \sum_{n=1}^{M}\{y_{nc}(t) + y_{ns}(t)\} \tag{4.66}$$

and the relevant boundary conditions [eqns. (4.59) and (4.60)].

The truncated model also gives some insight into the structure of the system transfer-function matrix[12] $G(s)$ relating the output $y(s)$ to the input $u(s)$. Namely, if $G_0(s)$, $G_{nc}(s)$ and $G_{ns}(s)$ are the transfer function matrices relating y_0, y_{nc} and y_{ns}, respectively, to the input $u(t)$, then a simple calculation yields the identity

$$G(s) = G_0(s) + \sum_{n=1}^{M}\{G_{nc}(s) + G_{ns}(s)\} \tag{4.67}$$

Lumped-parameter state-space models of eqns. (4.65) and (4.66) could be derived by the use of finite-difference approximations to spatial derivatives or by modal techniques analogous to those described earlier in the chapter. The models have the natural 'block diagonal' form,

$$\mu_e \dot{x}_0(t) = A_0 x_0(t) + B_0 u(t) \qquad y_0(t) = C_0 x_0(t)$$

$$\mu_e \dot{x}_{nc}(t) = A_{nc} x_{nc}(t) + B_{nc} u(t) \qquad y_{nc}(t) = C_{nc} x_{nc}(t) \qquad n \ge 1 \tag{4.68}$$

$$\mu_e \dot{x}_{ns}(t) = A_{ns} x_{ns}(t) + B_{ns} u(t) \qquad y_{ns}(t) = C_{ns} x_{ns}(t) \qquad n \ge 1$$

yielding the transfer-function matrix model

$$G(s) = C_0(s\mu_e - A_0)^{-1}B_0$$

$$+ \sum_{n=1}^{M}\{C_{nc}(s\mu_e - A_{nc})^{-1}B_{nc} + C_{ns}(s\mu_e - A_{ns})^{-1}B_{ns}\} \tag{4.69}$$

Remembering that the partial differential equations of (4.65) are defined in the two-dimensional (r, z) rather than the original three-dimensional (r, θ, z) reactor volume, the numerical advantages are apparent. For example, if the (r, z) plane is divided into 150 mesh areas, each block model of (4.68) will have dimension 450 (still large but much better than the dimensions of three-dimensional finite-difference models). The use of modal series expansion approximations will reduce these dimensions considerably more although. in practice, a combination of the techniques is probably most efficient.[10]

4.8 Combined modal and finite difference models[6, 10]

A combined modal and finite-difference scheme for approximating eqns. (4.65) can be derived based on radial zoning of the (r, z) plane shown in Fig. 4.6. It is motivated by the need to produce accurate lumped-parameter approximations of manageable dimension for numerical calculations. It is based on the observation[6, 10] that radial dynamics are more important than axial dynamics in reactor stability and control studies and hence that it is possible to use fairly crude approximations to axial spatial dynamics. More

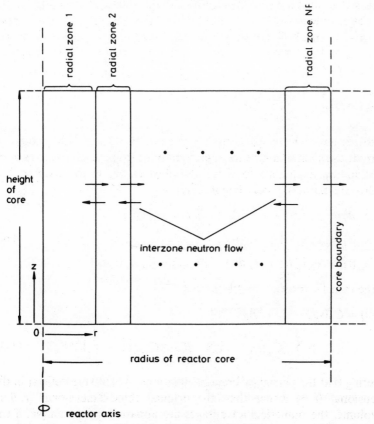

Fig. 4.6 *Radial zoning system*

precisely, it is assumed that, in radial zone j $(1 \leq j \leq NR)$, the vectors of interest can be expressed as a finite linear combination of spatially separable forms. Taking, for illustrative purposes, the case of ψ_0, we write

$$\psi_0(r, z, t) = \sum_{k=1}^{NA} D_{jk}(z)\Phi_{jk}(r, t) \tag{4.70}$$

in radial zone j. The 3×3 matrices $D_{jk}(z)$ are all diagonal matrices of known axial 'synthesis modes'. Without loss of generality, we can assume that they are orthonormal in the sense that

$$\int_0^H D_{jk}(z)D_{ji}(z)\, dz = \delta_{ki}I_3 \qquad 1 \leq i \qquad k \leq NA \tag{4.71}$$

The terms $\Phi_{jk}(r, t)$, $1 \leq k \leq NA$, are unknown vector functions of radial position and time.

Substituting into (4.65), multiplying by $D_{ji}(z)$ and integrating over the interval $0 \leq z \leq H$ will eliminate the axial dependence from the equations. Repeating this for $1 \leq i \leq NA$ and for each radial zone yields the model

$$\mu \frac{\partial \Phi_{ji}}{\partial t}(r, t) = \sum_{k=1}^{NA} L_0^{jik}\Phi_{jk}(r, t) + F_0^{ji}(r)u(t)$$

$$1 \leq i \leq NA \qquad 1 \leq j \leq NR \tag{4.72}$$

where

$$L_0^{jik} \triangleq \int_0^H D_{ji}(z)L_0D_{jk}(z)\, dz$$

is a 3×3 spatial operator involving radial derivatives only and radial dependent coefficients, and

$$F_0^{ji}(r) \triangleq \int_0^H D_{ji}(z)F_0\, dz \tag{4.73}$$

Typically, good accuracy in stability and frequency response calculations can be obtained with $NA = 1$ or 2 so that a significant reduction in the dimension of the representation of axial dynamics is possible compared with direct finite-difference methods.

Defining the spatial average over zone j as

$$\Phi_{ji}(t) \triangleq \frac{\int_{\text{zone } j} \Phi_{ji}(r, t)r\, dr}{\int_{\text{zone } j} r\, dr} \tag{4.74}$$

averaging both sides of eqn. (4.72) in the same way and approximating the radial derivatives by finite difference methods[6, 10] in terms of the $\Phi_{ji}(t)$, leads, after much manipulation, to a state space model of dimension $N = (3NA)NR$. Typically $NA = 2$ and $NR = 15$ leading to a model of dimension 90. This is large but manageable as the matrices involved tend not to be ill-conditioned. If $NA = 1$, the dimension reduces to 45 which is easily coped with.

Finally, the choice of axial synthesis modes requires careful thought[6, 10, 13] and should be based on the purposes for which the model was constructed. On the assumption that the model is required for stability and control studies, techniques can be derived based on the use of steady state data, a little guesswork and some involved calculations aimed at minimizing the errors in estimating the xenon feedback term in eqn. (4.18). The interested reader is referred to the references for more detail.

4.9 Summary

In many ways the modelling exercise in the analysis of spatial kinetics in nuclear reactor systems is an exercise in physical approximation and numerical reduction of the complex nonlinear partial differential equations describing the space-time behaviour of the reactor power distribution within the core. This is particularly important in the area of stability and control studies using eigenvalue,[5, 8, 9, 10] frequency domain[6, 10, 12, 14, 15] and optimization methods[4, 5, 6, 8] which, for numerical feasibility, require linear-state space models of relatively low dimension. There are many ways[5] of achieving this objective depending upon the accuracy required and the available data. This chapter has outlined, in the context of the modelling of zenon-induced oscillations in thermal reactor systems, how the ideas of modal expansion and finite difference methods can be used together to provide a successful solution to the problem.

4.10 Acknowledgment

Thanks are due to Professor H. Nicholson for providing the opportunity to publish this work and to the United Kingdom Atomic Energy Authority, Atomic Energy Establishment, Winfrith, for granting permission to draw on material from unclassified AEEW reports.

4.11 References

1 SYRETT, J. J.: 'Nuclear reactor theory', Nuclear Engineering Monographs (Temple Press, London, 1959)
2 HITCHCOCK, A.: 'Nuclear reactor stability' (ibid, 1960)
3 SCHULTZ, M. A.: 'Control of nuclear reactors and power plants' (McGraw-Hill, 1955)
4 WEAVER, L. E.: 'Reactor dynamics and control', American Elsevier Publishing Company, New York, 1968.

5 STACEY, W. M., Jnr.: 'Space-time nuclear reactor kinetics', Nuclear Science and Engineering Monograph 5 (Academic Press, 1969)
6 OWENS, D. H.: 'Multivariable control analysis of distributed parameter nuclear reactors', Ph.D. Thesis, Imperial College, 1973
7 KAPLAN, S.: 'The property of finality and the analysis of problems in reactor space-time kinetics by various modal expansions', *Nucl. Sci. Eng.*, 1961, **9**, pp. 357–361
8 WIBERG, D. M.: 'Optimal control of nuclear reactor systems', Advances in control systems 5 (Academic Press, 1967)
9 SUMNER, H. M.: 'ZIP Mk 2: A FORTRAN code for calculating the eigenvalues of large sets of linear equations', AEEW-R 543, 1969, HMSO
10 OWENS, D. H.: 'XENFER: A FORTRAN code for the calculation of spatial transfer functions of thermal reactors subject to xenon poison effects", AEEW-R 816, 1973, HMSO
11 HALMOS, P. R.: 'Finite-dimensional vector spaces' (Van Nostrand, 1958)
12 OWENS, D. H.: 'Feedback and multivariable systems', Control Engineering Series 7 (Peter Peregrinus, 1978)
13 OWENS, D. H.: 'Calculation of nuclear reactor spatial transfer functions using the computer programme XENFER', AEEW-R 817, 1973, HMSO
14 OWENS, D. H.: 'Multivariable-control-systems design concepts in the failure analysis of a class of nuclear reactor spatial control systems', *Proc. IEE*, 1973, **120**(1), pp. 119–125
15 OWENS, D. H.: 'Dyadic approximation method for multivariable control-systems design with a nuclear reactor application', ibid, pp. 801–809
16 BENNET, D. J.: 'The elements of nuclear power' (Longmans, 1972)

Chapter 5

Aerospace systems

J. M. Lipscombe

List of principal symbols

A	state matrix
a_x, a_y, a_z	co-ordinates of centre of gravity (Section 5.3)
a_x, a_y, a_z	accelerations in platform axes (Section 5.7)
a'_x, a'_y, a'_z	corrected accelerations (Section 5.7)
B	input matrix
b_x, b_y, b_z	accelerometer co-ordinates
C	output matrix
C^*	aircraft handling parameter (Section 5.6.3)
c_x, c_y, c_z	engine thrust moment arms (eqn. 5.9)
D	direct coupling matrix
E, E_x, E_y, E_z	engine thrust and its components
e	specific engine thrust, E/m
e_0, e_1, e_2, e_3	attitude parameters (Section 5.7.4)
f_L	line of sight acceleration
f_T	target acceleration
f_m	missile lateral acceleration (latax)
f_x, f_y, f_z	components of vehicle acceleration
$\lvert f_{max} \rvert$	maximum latax
G	error covariance matrix (Section 5.7.5)
g	acceleration due to gravity
g_x, g_y, g_z	components of g (Section 5.7)
H	nominal height (Section 5.6.1)
h	height
h_a	height measured by air data system
$\hat{h}$	best estimate of height

I	unit matrix
I_x, I_y, I_z	moments of inertia about axes through c.g.
K	state feedback matrix
K	constant in human pilot model (Section 5.6.4)
k	proportional navigation gain (Section 5.5.4)
k	constant defined by eqn. 5.63 (Section 5.6.1)
k_a	accelerometer feedback gain
k_r	rate gyroscope feedback gain
k_1, k_2	components of K
L	moment about Ox (Fig. 5.1)
L_a	moment about Ox due to aerodynamic effects
L_e	moment about Ox due to engine thrust
L_g	moment about Ox due to gravitation
L_p	$\partial L/\partial p$
L_r	$\partial L/\partial r$
L_v	$\partial L/\partial v$
L_ζ	$\partial L/\partial \zeta$ $\quad$ aerodynamic derivatives for rolling moment
L_ξ	$\partial L/\partial \xi$
$L_{\dot p}$	$\partial L/\partial \dot p$
$L_{\dot r}$	$\partial L/\partial \dot r$
l_p	$\doteq L_p/I_x$
l_r	$\doteq L_r/I_x$
l_v	$\doteq L_v/I_x$ $\quad$ for exact values see Section 5.6.2
l_ζ	$\doteq L_\zeta/I_x$
l_ξ	$\doteq L_\xi/I_x$
M	moment about Oy (Fig. 5.1)
M	mach number (Fig. 5.3)
M_a	moment about Oy due to aerodynamic effects
M_e	moment about Oy due to engine thrust
M_g	moment about Oy due to gravitation
M_q	$\partial M/\partial q$
M_u	$\partial M/\partial u$
M_w	$\partial M/\partial w$ $\quad$ aerodynaimic derivatives for pitching moment
M_η	$\partial M/\partial \eta$
$M_{\dot q}$	$\partial M/\partial \dot q$
$M_{\dot w}$	$\partial M/\partial \dot w$
m	vehicle mass
m_e	$\doteq mc_z/I_y$
m_q	$\doteq M_q/I_y$
m_u	$\doteq M_u/I_y$ $\quad$ for exact values see Section 5.6.1
m_w	$\doteq M_w/I_y$
N	moment about Oz (Fig. 5.1)
N_a	moment about Oz due to aerodynamic effects
N_e	moment about Oz due to engine thrust
N_g	moment about Oz due to gravitation

N_p	$\partial N/\partial p$
N_r	$\partial N/\partial r$
N_v	$\partial N/\partial v$
N_ζ	$\partial N/\partial \zeta$ \} aerodynamic derivatives for yawing moment
N_ξ	$\partial N/\partial \xi$
$N_{\dot p}$	$\partial N/\partial \dot p$
$N_{\dot r}$	$\partial N/\partial \dot r$
n_p	$\doteq N_p/I_z$
n_r	$\doteq N_r/I_z$
n_v	$\doteq N_v/I_z$ \} for exact values see Section 5.6.2
n_ζ	$\doteq N_\zeta/I_z$
n_ξ	$\doteq N_\xi/I_z$
P	angular velocity about Ox (Fig. 5.1)
P	Kalman filter gain matrix (Section 5.7.5)
$P(s)$	transfer function model of human pilot (Section 5.6.4)
p	perturbation angular velocity about Ox
p_1, p_2	elements of Kalman filter gain matrix, P
Q	angular velocity about Oy (Fig. 5.1)
$Q(s)$	transfer function of pilot plus plant (Fig. 5.11)
q	perturbation angular velocity about Oy
q	$q\delta(t) = \epsilon\alpha^2$ (Section 5.7.5)
R	angular velocity about Oz (Fig. 5.1)
R	radius of the earth (Section 5.7)
R_m	missile range from tracker (Fig. 5.5)
R_r	distance of missile from target (Fig. 5.7)
r	perturbation angular velocity about Oz
r	$r\delta(t) = \epsilon\beta^2$ (Section 5.7.5)
s	Laplace transform variable
T	time of missile-target impact
t	time variable
U	velocity along Ox (Fig. 5.1)
U_G	gust velocity along Ox
U_T	target velocity
U_m	missile velocity
U_r	relative velocity of missile and target
U_0	velocity along $O_0 x_0$
U_{co}	cross-over velocity (Section 6.6.3)
U_{0G}	gust velocity along $O_0 x_0$
u	perturbation velocity along Ox
u	input vector
V	velocity along Oy (Fig. 5.1)
V_G	gust velocity along Oy
V_T	$(U^2 + V^2 + W^2)^{1/2}$
V_x, V_y, V_z	velocities with respect to a tangent plane rotating with the earth (Section 5.7.1)

V_0	velocity along $O_0 y_0$
V_{xe}, V_{ye}, V_{ze}	velocity of the tangent plane due to the earth's rotation (Section 5.7.1)
V_{0G}	gust velocity along $O_0 y_0$
$\dot{V}_{zi}$	inertial platform measurement of vertical acceleration
v	perturbation velocity along Oy
W	velocity along Oz (Fig. 5.1)
W_G	gust velocity along Oz
W_0	velocity along $O_0 z_0$
W_{0G}	gust velocity along $O_0 z_0$
w	perturbation velocity along Oz
X	force along Ox (Fig. 5.1)
X_a	force along Ox due to aerodynamic effects
X_e	force along Ox due to engine thrust
X_g	force along Ox due to gravitation
X_u	$\partial X/\partial u$
X_w	$\partial X/\partial w$ aerodynamic derivatives for longitudinal force
X_η	$\partial X/\partial \eta$
x_u	$\doteq X_u/m$
x_w	$\doteq X_w/m$ for exact values see Section 5.6.1
x_η	$\doteq X_\eta/m$
x_1, x_2	components of $\mathbf{x}$
$\mathbf{x}$	state vector
$\hat{x}_1, \hat{x}_2$	components of $\hat{\mathbf{x}}$
$\hat{\mathbf{x}}$	best estimate of $\mathbf{x}$
Y	force along Oy (Fig. 5.1)
Y_a	force along Oy due to aerodynamic effects
Y_e	force along Oy due to engine thrust
Y_g	force along Oy due to gravitation
Y_r	$\partial Y/\partial r$
Y_v	$\partial Y/\partial v$ aerodynamic derivatives for lateral force
Y_ζ	$\partial Y/\partial \zeta$
y_v	$\doteq Y_v/m$
y_ζ	$\doteq Y_\zeta/m$ for exact values see Section 5.6.2
$\mathbf{y}$	output vector
Z	force along Oz (Fig. 5.1)
Z_a	force along Oz due to aerodynamic effects
Z_e	force along Oz due to engine thrust
Z_g	force along Oz due to gravitation
Z_q	$\partial Z/\partial q$
Z_u	$\partial Z/\partial u$
Z_w	$\partial Z/\partial w$ aerodynamic derivatives for normal force
Z_η	$\partial Z/\partial \eta$
$Z_{\dot{q}}$	$\partial Z/\partial \dot{q}$
$Z_{\dot{w}}$	$\partial Z/\partial \dot{w}$

z_e	$\doteq \epsilon$
z_q	$\doteq Z_q/m$
z_u	$\doteq Z_u/m$ } for exact values see Section 5.6.1
z_w	$\doteq Z_w/m$
z_η	$\doteq Z_\eta/m$
A	angle of attack
α	perturbation angle of attack
α	acceleration constraint constant (Section 5.5.4)
α	phase advance constant (Section 5.6.4)
α	inertial platform error (Section 5.7.5)
B	angle of sideslip
β	perturbation angle of slideslip
β	air data computer error (Section 5.7.5)
γ	nominal value of ϕ_m
$\delta(t)$	unit impulse function
$\Delta y, \Delta\theta, \Delta\psi$	defined in Fig. 5.5 (Section 5.5.2)
$\Delta f_x, \Delta h, \Delta u$	
$\Delta V_y, \Delta x, \Delta \beta$	errors associated with inertial navigation unit (Section 5.7)
$\Delta\theta, \Delta\omega_x$	
$\delta t, \delta y, \delta\theta$	defined in Fig. 5.9 (Section 5.5.4)
$\delta\phi_m$	
ϵ	E_z/E
ζ	rudder deflection
ζ_c	rudder command from guidance system (Section 5.5)
η	elevator deflection
Θ	pitch angle
θ	perturbation pitch angle
θ	angle of line of sight (Section 5.5)
Λ	longitude
λ	latitude
μ	damping ratio
μ_d	damping ratio of dutch roll mode
μ_p	damping ratio of phugoid mode
μ_q	damping ration of short period mode
ξ	aileron deflection
ρ	air density
τ_r	time constant of roll subsidence mode
τ_s	time constant of spiral mode
τ_a, τ_d, τ_e	time constants of pilot behaviour (Section 5.6.4)
Φ	roll angle
ϕ	perturbation roll angle
ϕ_T	target heading angle (Section 5.5.4)
ϕ_m	missile heading angle (Section 5.5.4)
Ψ	azimuth angle
ψ	perturbation azimuth angle

Ω	rotational speed of the earth
ω_e	bandwidth associated with human pilot model (Section 6.4)
ω_d	natural frequency of dutch roll mode
ω_n	natural frequency
ω_p	natural frequency of phugoid mode
ω_q	natural frequency of short period mode
$\omega_x, \omega_y, \omega_z$	inertial platform rotation rates
ω_{xe}	drift error in x axis integrating rate gyroscope
Ω_{xe}	a constant value of ω_{xe}

Mathematical notation

$\dot{x}$	derivative of x
$\mathbf{x}$	column matrix col $[x_1, x_2, x_3, ..., x_m]$
$\bar{x}$	Laplace transform of x
∂E	perturbation of E
A^T	the transpose of A
ϵw	expectation of w
det A	determinant of A
$\Rightarrow$	implies

Abbreviations

c.g.	centre of gravity
DCM	direction cosine matrix
DLC	direct lift control
DSFC	direct side force control
INU	inertial navigation unit
IRG	integrating rate gyroscope
latax	lateral acceleration at missile c.g.
LOS	line of sight
SAS	stability augmentation system
SL	sight line

5.1 Introduction

Traditionally the modelling of aerospace systems has fallen into two parts

(a) three-dimensional geometry and kinematics
(b) the dynamics of a rigid body and the fluid (air) through which it moves.

Developments through the past few decades have forced the attention of aerospace engineers toward other models such as

(c) structural elasticity (structural mode controllers) and the interaction with the fluid (flutter suppression controllers, gust alleviation systems, etc.)
(d) models of the propulsion system

(*e*) dynamics of the hydraulic or other actuators which power the aerodynamic control surfaces

(*f*) the behaviour of the human pilot in response to natural and artificial cues

(*g*) the navigation system (this largely applies to aircraft and covers an enormous range of disciplines from the errors in inertial navigation units to the radio wave propagation problems of instrument landing systems)

(*h*) guidance problems (this is roughly the missile equivalent of (*g*), and includes models as diverse as laser beam riding for antitank weapons and minimum-fuel rendezvous trajectories for space vehicles).

The above models are only a selection of those currently used by aerospace engineers, and ignore the control system itself; typically (in an aircraft) this consists of a large number of redundant sensors and separated digital computing systems which not only have complex internal dynamics, but give rise to difficult and interesting modelling problems in terms of reliability, maintainability, cost of ownership and overall system integrity. This chapter concentrates on models (*a*) and (*b*), but inevitably the characteristics of some of the other models intrude.

The layout of the chapter is as follows. The basic equations of vehicle geometry and dynamics are presented in Sections 5.2 and 5.3, respectively. They are reduced to a set of linear perturbation equations in Section 5.4. Three examples of their use in modelling various aerospace systems are given in the remaining three sections. The lateral equations of motion of a line-of-sight guided missile are given in Section 5.5, the equations of motion of a conventional aircraft in Section 5.6, and the behaviour of an inertial navigation unit in Section 5.7.

5.2 Geometry of aerospace vehicles

This model and its various simplifications describe the position and attitude of the vehicle with respect to a set of reference axes. For the purposes of evaluating the aircraft performance, the reference axes are stationary with respect to the fixed stars (i.e. are 'inertial axes'), but when navigation systems are modelled a third set describing navigation co-ordinates on a spherical rotating earth must be invoked (Section 5.7.1).

5.2.1 Definitions

Body axes: A right-handed set of orthogonal axes fixed to the aircraft as shown in Fig. 5.1

Ox points 'forward' (*longitudinal* axis)

Oy points to starboard (*lateral* axis)

Oz points 'down' (*normal* axis)

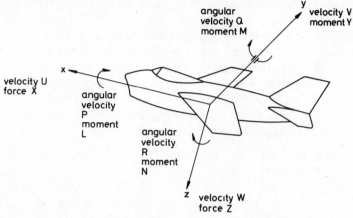

Fig. 5.1   *Diagram of body axes and velocities*

Reference axes: A right-handed set of orthogonal axes in the tangent plane to the earth's surface

$O_0 x_0$ points east (or sometimes north)

$O_0 y_0$ points south (if $O_0 x_0$ points north, $O_0 y_0$ must point east)

$O_0 z_0$ points towards the centre of the earth.

Body velocities: The aircraft has velocities

U along Ox

V along Oy

W along Oz

and angular velocities

P about Ox (*roll* rate)

Q about Oy (*pitch* rate)

R about Oz (*yaw* rate)

The angular velocities are right handed, i.e. clockwise when looking in the direction of the appropriate axis.

Velocity with respect to the reference axes: Let the vehicle have velocity

U_0 along $O_0 x_0$

V_0 along $O_0 y_0$

W_0 along $O_0 z_0$

Note that $W_0 = -dh/dt$ where h is the *height* of the aircraft.

Aircraft attitude: The orientation of $Oxyz$ with respect to $O_0 x_0 y_0 z_0$ is defined by the following sequence of rotations

(i) let an intermediate axis set $O_i x_i y_i z_i$ be coincident with $O_0 x_0 y_0 z_0$.

(ii) rotate $O_i x_i y_i z_i$ about the axis Oz_i by an angle Ψ (*azimuth* angle)

(iii) rotate $O_i x_i y_i z_i$ about the axis Oy_i by an angle Θ (*pitch* angle)

(iv) rotate $O_i x_i y_i z_i$ about the axis Ox_i by an angle Φ (*roll* angle)

(v) these rotations have been made in such a way that $O_i x_i y_i z_i$ is now coincident with $Oxyz$.

Then Ψ, Θ, Φ define the attitude of $Oxyz$ with respect to $O_0x_0y_0z_0$ and are called the *Euler* angles of the aircraft. They are positive if the rotations are right handed.

5.2.2 Relationships between body and reference axes
It can be shown[1] that

$$\begin{bmatrix} U_0 \\ V_0 \\ W_0 \end{bmatrix} = \begin{bmatrix} c\Theta c\Psi & s\Phi s\Theta c\Psi - c\Phi s\Psi & c\Phi s\Theta c\Psi + s\Phi s\Psi \\ c\Theta s\Psi & s\Phi s\Theta s\Psi + c\Phi c\Psi & c\Phi s\Theta s\Psi - s\Phi s\Psi \\ -s\Theta & s\Phi c\Theta & s\Phi s\Theta \end{bmatrix} \begin{bmatrix} U \\ V \\ W \end{bmatrix} \tag{5.1}$$

where c stands for cosine and s for sine

Clearly the 3×3 matrix in eqn. (5.1) is a *direction cosine matrix* (DCM) written in terms of the Euler angles instead of the more usual form (in applied mathematics texts) which uses the direction cosines between $Oxyz$ and $O_0x_0y_0z_0$.

The attitude of the aircraft changes in response to its roll, pitch and yaw velocities. It is simple to show[1] that the relationship is

$$\begin{bmatrix} \dot\Phi \\ \dot\Theta \\ \dot\Psi \end{bmatrix} = \begin{bmatrix} 1 & \sin\Phi\tan\Theta & \cos\Phi\tan\Theta \\ 0 & \cos\Phi & -\sin\Phi \\ 0 & \sin\Phi\sec\Theta & \cos\Phi\sec\Theta \end{bmatrix} \begin{bmatrix} P \\ Q \\ R \end{bmatrix} \tag{5.2}$$

Equation (5.2) indicates one of the main disadvantages of representing attitude by Euler angles: the equation breaks down at $\Theta = \pm 90°$, and therefore the description is of limited use for simulation purposes, (see eqns. (5.88), (5.89) and (5.90) for a suitable alternative). However, the Euler angles are convenient for analytical work because they are easy to visualize. For example, $[\Phi\Theta\Psi] = [5° \; 30° \; 45°]$ immediately conveys the information that the aircraft is climbing at $30°$, rolled slightly to starboard and on a heading $45°$ to the reference direction O_0x_0.

The inverse of eqn. (5.2) is

$$\begin{bmatrix} P \\ Q \\ R \end{bmatrix} = \begin{bmatrix} 1 & 0 & -\sin\Theta \\ 0 & \cos\Phi & \sin\Phi\cos\Theta \\ 0 & -\sin\Phi & \cos\Phi\cos\Theta \end{bmatrix} \begin{bmatrix} \dot\Phi \\ \dot\Theta \\ \dot\Psi \end{bmatrix} \tag{5.3}$$

Notice that P, Q Φ and Θ have been named in a rather unfortunate way because the pitch and roll angles (Θ and Φ) are not in general the integral of the pitch and roll rates (Q and P).

Figure 5.2 shows how eqns. (5.1) and (5.2) transform body velocities into attitude and velocity with respect to the reference axes.

5.2.3 Angles of attack and sideslip
Before leaving geometrical ideas, it is possible to transform U, V and W into total aircraft velocity, and the angles of attack and sideslip, which are perhaps easier to visualize.

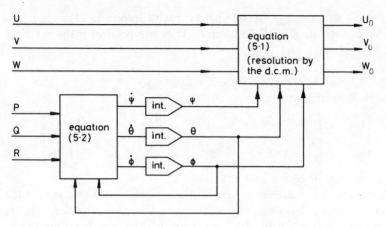

Fig. 5.2   *Rigid body geometry*

Define the *total velocity* of the aircraft

$$V_T = (U^2 + V^2 + W^2)^{1/2}$$

Then the *angle of attack* is defined by

$$\sin A = \frac{W}{(U^2 + W^2)^{1/2}}$$

and the *angle of sideslip* is defined by

$$\sin B = \frac{V}{V_T}$$

Thus

$$U = V_T \cos A \cos B$$
$$V = V_T \sin B \tag{5.4}$$
$$W = V_T \sin A \cos B$$

5.3 Dynamics

It is conventional[1,2] to start with the so-called *Euler equations* of a rigid body moving in a vacuum under the influence of external forces. In fact these equations are in error because they do not properly account for the mass of the displaced fluid. However, for aircraft and missiles this effect is quite small and can be approximated by adding appropriate external forces to the right-hand side of the equations. The models of submarines or airships may sometimes be formulated with more complicated dynamics.

5.3.1 Euler equations

$$[\dot{U} + WQ - RV - a_x(R^2 + Q^2) + a_y(PQ - \dot{R}) + a_z(PR + \dot{Q})]m$$
$$= X$$

$$[\dot{V} + UR - PW + a_x(PQ + \dot{R}) - a_y(P^2 + R^2) + a_z(RQ - \dot{P})]m$$
$$= Y$$

$$[\dot{W} + VP - QU + a_x(RP - \dot{Q}) + a_y(RQ + \dot{P}) - a_z(Q^2 + P^2)]m$$
$$= Z$$

$$I_x\dot{P} + (I_z - I_y)RQ - I_{yz}(Q^2 - R^2) - I_{xz}(\dot{R} + PQ) - I_{xy}(\dot{Q} - PR)$$
$$= L + Ya_z - Za_y$$

$$I_y\dot{Q} + (I_x - I_z)PR - I_{yz}(\dot{R} - PQ) - I_{xz}(R^2 - P^2) - I_{xy}(\dot{P} + QR)$$
$$= M + Za_x - Xa_z$$

$$I_z\dot{R} + (I_y - I_x)QP - I_{yz}(\dot{Q} + RP) - I_{xz}(\dot{P} - RQ) - I_{xy}(P^2 - Q^2)$$
$$= N + Xa_y - Ya_x$$

$$(5.5)$$

where m is the mass of the aircraft, I_x, I_y, I_z, I_{xy}, I_{yz}, I_{xz} are the moments of inertia about axes through the centre of gravity (c.g.) but parallel to Ox, Oy and Oz, X, Y and Z are the forces along Ox, Oy and Oz, L, M and N are the moments about Ox Oy and Oz (in a right-handed sense) and a_x, a_y and a_z are the co-ordinates of the aircraft centre of gravity with respect to $Oxyz$.

For these equations to be valid U, V, W, P, Q and R must be absolute velocities. Application of eqns. (5.1) and (5.2) therefore result in U_0, V_0, W_0, Φ, $\dot{\Theta}$ and $\dot{\Psi}$ which are with respect to inertial axes, i.e. a fixed flat earth. However, for aircraft or missile performance modelling this is quite satisfactory. The navigational model is a different matter, and this will be discussed in Section 5.7.

Equation (5.5) simplifies if the origin of the body axes can be placed at the c.g. $(a_x = a_y = a_z = 0)$, and the assumption made that the axes are principal ones $(I_{xy} = I_{yz} = I_{zx} = 0)$. The Euler equations then reduce to

$$m(\dot{U} + WQ - RV) = X$$
$$m(\dot{V} + UR - PW) = Y$$
$$m(\dot{W} + VP - QU) = Z$$
$$I_x\dot{P} + (I_z - I_y)RQ = L$$
$$I_y\dot{Q} + (I_x - I_z)PR = M$$
$$I_z\dot{R} + (I_y - I_x)QP = N$$

$$(5.6)$$

5.3.2 Aircraft acceleration

In terms of the body velocities (U, V, W, P, Q, R), the acceleration experienced in the aircraft at a point with co-ordinates $(b_x b_y b_z)$ with respect to $Oxyz$ is given by[2]

$$
\begin{aligned}
f_x &= \dot{U} + WQ - RV - b_x(R^2 + Q^2) + b_y(PQ - \dot{R}) \\
&\quad + b_z(PR + \dot{Q}) + g \sin \Theta \\
f_y &= \dot{V} + UR - PW + b_x(PQ + \dot{R}) - b_y(P^2 + R^2) \\
&\quad + b_z(RQ - \dot{P}) - g \cos \Theta \sin \Phi \\
f_z &= \dot{W} + VP - QU + b_x(RP - \dot{Q}) + b_y(RQ + \dot{P}) \\
&\quad - b_z(Q^2 + P^2) - g \cos \Theta \cos \Phi
\end{aligned}
\tag{5.7}
$$

where f_x, f_y and f_z are the accelerations along Ox, Oy and Ox, respectively. The reader will find it instructive to compare eqn. (5.7) with the first three equations of (5.5), and put a physical meaning to each term.

5.3.3 Gravitional forces

Equation (5.6) (or in awkward cases (5.5)) transforms a set of forces and moments acting on the aircraft into body velocities. These can then be transformed by eqns. (5.1) and (5.2) into vehicle motion with respect to the earth. Clearly the next task is to build up a model to generate the forces X, Y and Z and the moments L, M and N.

First consider the forces and moments which are due to gravitational attraction. Simple geometry shows that

$$
\begin{bmatrix} X_g \\ Y_g \\ Z_g \\ L_g \\ M_g \\ N_g \end{bmatrix} =
\begin{bmatrix}
-\sin \Theta \\
\cos \Theta \sin \Phi \\
\cos \Theta \cos \Phi \\
a_y \cos \Theta \cos \Phi - a_z \cos \Theta \sin \Phi \\
-a_z \sin \Theta - a_x \cos \Theta \cos \Phi \\
a_y \sin \Theta + a_x \cos \Theta \sin \Phi
\end{bmatrix} mg
\tag{5.8}
$$

where the subscript g means 'due to gravitation'.

If the origin of the body axes is placed at the c.g., $a_x = a_y = a_z = 0$ and the three gravitational moments become zero.

5.3.4 Aerodynamic forces

It is assumed that each force is a function of

(i) air density, ρ
(ii) relative motion of the vehicle through the air, and
(iii) the vehicle configuration.

Consider each of these in turn. Clearly the density of the air (which is mainly a function of height, h) influences the magnitude of any aerodynamically generated force. The variable h forms the ordinate of the flight envelope (Fig. 5.3) which is described in Section 5.4.1.

The relative motion of the vehicle through the air is not only a function of the body velocities, but also a function of any movement of the air relative to the reference axes. Let these *gust velocities* be U_{OG}, V_{OG} and W_{OG}. Then transformation through the direction cosine matrix (DCM) by using the inverse of eqn. (5.1) gives the gust velocities resolved along the aircraft body axes, U_G, V_G and W_G.

A reasonable assumption might be that each aerodynamic force or moment is a function of $(U, V, W, P, Q, R, \dot{U}, \dot{V}, \dot{W}, \dot{P}, \dot{Q}, \dot{R}, U_G, V_G, W_G, \dot{U}_G, \dot{V}_G, \dot{W}_G, \rho)$. One of the main reasons for reformulating the equations in terms of perturbation velocities (Section 5.4) is to reduce this function to one which can be handled analytically. Even simulation studies usually use simplified semi-analytical approximations.

The word configuration is usually used to denote a 'steady-state' aircraft shape, i.e. whether it has the undercarriage extended or not, or the position of the landing flaps. In this section, however, the meaning is extended to include control surface deflections. There are conventionally three aerodynamic controls on an aircraft, each designed to produce a moment about one of the three axes. The *ailerons* deflect by an angle ζ to produce a moment about Ox; the *elevator* deflects by an angle η to produce a moment about Oy; and the *rudder* deflects by an angle ζ to produce a moment about Oz. In each case a *positive* deflection of the control surface produces a *negative* moment about the appropriate axis.

In traditional aircraft the ailerons, elevator and rudder are hinged flaps placed at the wing tips and tail. On most modern aircraft the ailerons are supplemented by additional aerodynamic surfaces on the wing. Indeed on

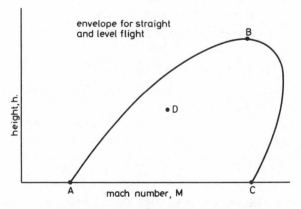

Fig. 5.3   *Sketch of a flight envelope*

some high performance aircraft every square inch of the aerodynamic surfaces seem movable for some control purpose. A selection of the control functions are:

(i) additional surfaces on the wing to alter its lift without creating a moment (direct lift control: DLC)
(ii) additional surfaces (which may work in conjunction with the rudder) to produce a lateral force without creating a moment (direct side force control: DSFC). These five aerodynamic controls, ξ, η, ζ, DLC and DSFC together with engine thrust and aerodynamic air-brakes to produce longitudinal forces give six control forces and moments to control the six degree of freedom dynamics expressed in equation (5.5).
(iii) Additional surfaces to resolve structural mode and aeroelasticity (flutter) problems. These surfaces may be necessary because in control theory language, the modes may be uncontrollable with respect to the main control surfaces. Additional instrumentation is often required to make them observable.
(iv) Variable sweep-back or flap deflection as an automatic function of aircraft manoeuvre. This extends the flight envelope without adding to the pilot's workload, and can improve the combat manoeuvrability.

5.3.5 Propulsion forces
An aircraft engine is a complicated machine which is as challenging to model as the aircraft itself. Fortunately, when studying the behaviour of an aircraft, drastic simplifications to the engine model may usually be made without compromising the results.

Let the engine have thrust E with components E_x E_y and E_z along the axes Ox, Oy and Oz, and the perpendicular from the thrust line (i.e. the direction of E) to the axis origin have components c_x c_y and c_z.
Then

$$
\begin{aligned}
X_e &= E_x & L_e &= c_y E_z - c_z E_y \\
Y_e &= E_y & M_e &= c_z E_x - c_x E_z \\
Z_e &= E_z & N_e &= c_x E_y - c_y E_x
\end{aligned}
\tag{5.9}
$$

where the subscript e means 'due to engine thrust'.

5.3.6 Summary of Section 5.3
At this point, a respectable model of the aircraft motion has been constructed. Inputs in the form of gust velocities, aerodynamic configuration changes and engine commands modify the forces and moments acting on the aircraft. These, via eqn. (5.5) or (5.6), produce body velocities which can be translated into movement with respect to fixed axes by eqns. (5.1) and (5.2). A feedback loop exists from the body velocities via the aerodynamic functions (which have not yet been specified in a detailed form) to create aerodynamic forces,

X_a, Y_a, Z_a, L_a, M_a and N_a. Another feedback loop is created from the aircraft attitude Φ, Θ, Ψ via the gravitational forces by eqn. (5.8). The outputs of the model are, of course, the variables which are of interest for performance estimation, or which are easy to measure. Examples are angle of attack (definitions preceding eqn. (5.4)) and aircraft acceleration [eqn. (5.7)], respectively. It is an instructive exercise for the reader to complete Fig. 5.2 by adding the extra blocks and feedback loops.

The model is almost suitable for simulation. An analytical/look-up table description of the aerodynamic functions must be created for the particular aircraft under simulation, and perhaps some structural dynamics, actuator lags and engine characteristics added. (Where would these blocks appear in the extended version of Fig. 5.2?)

It is conventional to use the inverse of eqn. (5.4) to replace the variables U, V and W with V_T, A and B in many flight simulations for accuracy and bandwidth reasons.[3] This is a special case of the general mathematical modelling problem of transforming the state variables into the best form for processing through the computer. In the linear time-invariant case, for example, it often means putting the state matrix into companion form.

5.4 Perturbation equations

For analytical studies the above model is too complex. In this section perturbation methods are used to reduce the equations to linear time-invariant form suitable for control system design studies.

5.4.1 Flight envelope

A characteristic which aircraft and missiles share is that the total velocity V_T is predominantly composed of the forward velocity U. The aerodynamic characteristics of the aircraft vary greatly as the speed is changed. It has already been noted that the aerodynamics are a function of air density, ρ, and these two ideas lead to the concept of a *flight envelope* for a particular vehicle. Figure 5.3 shows a sketch of a flight envelope. The vertical scale is height h (representing ρ), and the horizontal scale is Mach number M (which is V_T divided by the ambient speed of sound). The interior of the envelope is the operational region of the aircraft. If it goes outside curve AB the wings are unable to generate enough lift to support the aircraft weight. Structural, thermal or power limitations prevent the aircraft crossing the boundary BC. The first step in forming the perturbation equations is to choose a point within the flight envelope around which to linearize. A full-scale study looks at many points and control laws are devised for each point. These are mechanized on the aircraft in such a way that the flight control system interpolates between them as the air data computer calculates where the aircraft is positioned in its envelope. In the next section the model will be formulated for just one point, D say.

5.4.2 Stability axes

If the body axes are oriented in the vehicle so that under nominal conditions U is the forward speed and $V = W = 0$, then the axes are called *stability* axes for that particular flight condition. Naturally the axes move with respect to the body as the aircraft changes its position in the flight envelope, but at a particular nominal operating condition the axes are fixed with respect to the aircraft. Assume steady horizontal flight so that $P = Q = R = \Phi = \Theta = \Psi = 0$. Small departures from the nominal condition give rise to perturbation velocities along the body-fixed stability axes of $[u, v, w]$ and angular velocities about them of $[p, q, r]$. The Euler angles are perturbed to $[\phi, \theta, \psi]$.

Neglecting second-order terms, eqn. (5.1) becomes

$$\begin{bmatrix} U_0 \\ V_0 \\ W_0 \end{bmatrix} = \begin{bmatrix} U \\ U\psi + v \\ -U\theta + w \end{bmatrix} \tag{5.10}$$

The last row is more usually written

$$\dot{h} = U\theta - w \tag{5.11}$$

Equation (5.2) becomes

$$\begin{bmatrix} \dot{\phi} \\ \dot{\theta} \\ \dot{\psi} \end{bmatrix} = \begin{bmatrix} p \\ q \\ r \end{bmatrix} \tag{5.12}$$

The three definitions of Section 5.2.3 reduce to

$$U = V_T$$

$$\alpha = \frac{w}{U} \quad \text{and} \quad \beta = \frac{v}{U} \tag{5.13}$$

where α and β are perturbation angles of attack and sideslip, respectively.

Stability axes are naturally placed at the centre of gravity, reducing eqns. (5.5) to

$$m\dot{u} = \partial X$$

$$m(\dot{v} + Ur) = \partial Y$$

$$m(\dot{w} - qU) = \partial Z$$

$$I_x \dot{p} - I_{xz}\dot{r} - I_{xy}\dot{q} = \partial L \tag{5.14}$$

$$I_y \dot{q} - I_{yz}\dot{r} - I_{xy}\dot{p} = \partial M$$

$$I_z \dot{r} - I_{yz}\dot{q} - I_{xz}\dot{p} = \partial N$$

where ∂ stands for 'the perturbation in'.

Sometimes the stability axes are far from being axes of symmetry and I_{xz} is not negligible, usually under low-speed conditions. Readers who have seen Concorde land, and recall that Ox is along the total unperturbed velocity vector, will appreciate the situation. Often, however, the cross products of inertia are relatively small, and eqn. (5.14) can be rewritten

$$m\dot{u} = \partial X$$
$$m(\dot{v} + Ur) = \partial Y$$
$$M(\dot{w} - qU) = \partial Z$$
$$I_x\dot{p} = \partial L \tag{5.15}$$
$$I_y\dot{q} = \partial M$$
$$I_z\dot{r} = \partial N$$

Equation (5.7) becomes

$$f_x = \dot{u} - b_y\dot{r} + b_z\dot{q} + g\theta$$
$$f_y = \dot{v} + Ur + b_x\dot{r} - b_z\dot{p} - g\phi \tag{5.16}$$
$$f_z = \dot{w} - Uq - b_x\dot{q} + b_y\dot{p}$$

where f_x, f_y and f_z are now perturbation accelerations. Equation (5.8) becomes (with $a_x = a_y = a_z = 0$)

$$\partial X_g = -mg\theta$$
$$\partial Y_g = mg\phi \tag{5.17}$$
$$\partial Z_g = \partial L_g = \partial M_g = \partial N_g = 0$$

Equation (5.9) becomes

$$\partial X_e = \partial E$$
$$\partial Z_e = \epsilon\,\partial E$$
$$\partial M_e = c_z\,\partial E \tag{5.18}$$
$$\partial Y_e = \partial L_e = \partial N_e = 0$$

where $\epsilon = E_z/E$, the thrust line inclination, and E_y, c_y and $E_z c_x$ are assumed negligible.

5.4.3 Aerodynamic derivatives
The aerodynamic force and moment functions mentioned in Section 5.3.4 can be expanded about their nominal value in a multidimensional Taylor series. For example, for a particular value of U and ρ (i.e. a particular point in the flight envelope)

$$\partial X_a = \frac{\partial X}{\partial u}.u + \frac{\partial X}{\partial v}.v + \cdots \frac{\partial X}{\partial \dot{q}}\dot{q} + \frac{\partial X}{\partial \dot{r}}\dot{r}$$

$$+ \frac{\partial X}{\partial \xi}\xi + \cdots + \text{higher order terms} \tag{5.19}$$

Note that configuration changes are included in the expansions [an aileron term $(\partial X/\partial \xi)\xi$ is shown]. The aileron, elevator and rudder deflections ξ, η and ζ must be regarded as perturbation deflections from the nominal values for that point in the flight envelope.

Those terms in expansion (5.19), which experiment or intuition indicate are negligible, are rejected and the remainder incorporated in the mathematical model. Under some extreme flight conditions certain second-order terms may be non-negligible, but this is unusual. For example, a typical aircraft expansion might be

$$\partial X_a = \frac{\partial X}{\partial u} u + \frac{\partial X}{\partial w} w + \frac{\partial X}{\partial \eta} \eta$$

It is conventional to write this

$$\partial X_a = X_u u + X_w w + X_\eta \eta$$

where X_u, X_w and X_η are called *aerodynamic derivatives*.

5.5 Model of the Lateral Equations of Motion of a Guided Missile

To illustrate the use of the equations given in Section 5.4, consider the model of lateral motion of the cruciform missile illustrated in Fig. 5.4. Control is via the deflection of the four rear fins, and it is assumed that the missile is stabilized in roll so that $\Phi = 0$. Further, assume that the pitch motion and yaw motion can be analysed separately because the symmetry of the missile eliminates aerodynamic coupling. (In fact this assumption breaks down at high angles of attack.) Geometrical coupling [eqn. (5.3)] is also ignored.

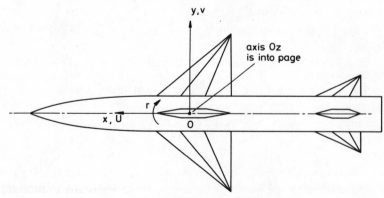

Fig. 5.4   *Diagram of a Cruciform Missile*

5.5.1 Lateral dynamics of the Missile

Consider each of the constituents of eqn. (5.15) in turn. The first is $m\dot{u} = \partial X$. If it is assumed that the missile performance is insensitive to small changes in forward speed, then the perturbation velocity u may be taken as zero. The first equation is therefore irrelevant and can be discarded.

The second equation, $m(\dot{v} + Ur) = \partial Y$, is the lateral force balance and is obviously of importance.

The third equation, $m(\dot{w} - qU) = \partial Z$, is the normal force balance and, by the initial assumptions, does not affect the lateral behaviour.

The fourth equation, $I_x \dot{p} = \partial L$, is the moment balance about the Ox axis. Because the missile is roll stabilized, the control system automatically adjusts ∂L by differential movement of the tail fins to keep $\Phi = 0$. Assume that the operation of this roll control loop does not influence the lateral motion of the missile.

The fifth equation is $I_y \dot{q} = \partial M$ and describes the pitching motion of the vehicle. By the initial assumptions, this is of no interest.

Finally the sixth equation, $I_z \dot{r} = \partial N$, which describes the moment balance about the Oz axis, is clearly of importance.

Thus the significant equations are

$$m(\dot{v} + Ur) = \partial Y$$
$$I_z \dot{r} = \partial N \tag{5.20}$$

The next stage in modelling the dynamics is to obtain expressions for ∂Y and ∂N. Examination of eqns. (5.17) and (5.18) reveals that $\partial N_g = \partial Y_e = \partial N_e = 0$ and $\partial Y_g = mg\phi$. Since $\phi = 0$ (roll stabilization), $\partial Y_g = 0$. Thus the only forces and moments are aerodynamic. Experience shows that the important aerodynamic derivatives are

Y_v Lateral force due to lateral velocity. Clearly this is always negative.

Y_r Lateral force due to yaw rate.

N_v Yaw moment due to lateral velocity. If the centre at which the aerodynamic forces act is behind the centre of gravity, then the missile 'turns into the wind' and N_v is positive. This is called *weathercock* or *static stability*. In principle N_v may be made positive by suitably positioning the wings and tail. However, both the centre of gravity and aerodynamic centre are subject to relatively large excursions during the missile flight.

N_r Yaw moment due to yaw rate. Clearly this is negative.

Y_ζ Lateral force due to rudder (rear fin) displacement.

N_ζ Yaw moment due to rudder displacement. The definition of positive rudder displacement is one which causes a negative moment. Thus N_ζ is always negative. The side force at the rear fin which is needed to produce this negative moment is in the direction of Oy, and therefore Y_ζ is positive.

Substituting these derivatives into eqn. (5.20) gives

$$m\dot{v} = Y_v v + (Y_r - mU)r + Y_\zeta \zeta$$
$$I_z \dot{r} = N_v v + N_r r + N_\zeta \zeta \qquad (5.21)$$

Because of the missile's high speed $|mU| \gg |Y_r|$. Thus Y_r may be neglected. (The same is not true for slow vehicles: for example most torpedos rely on Y_r to provide uncontrolled dynamic stability.)
Let

$$\frac{Y_v}{m} = y_v \qquad \frac{Y_\zeta}{m} = y_\zeta$$

$$\frac{N_v}{I_z} = n_v \qquad \frac{N_r}{I_z} = n_r \qquad \text{and} \qquad \frac{N_\zeta}{I_z} = n_\zeta$$

Equation (5.21) can then be written in the state form of $\dot{x} = Ax + Bu$:

$$\begin{bmatrix} \dot{v} \\ \dot{r} \end{bmatrix} = \begin{bmatrix} y_v & -U \\ n_v & n_r \end{bmatrix} \begin{bmatrix} v \\ r \end{bmatrix} + \begin{bmatrix} y_\zeta \\ n_\zeta \end{bmatrix} \zeta \qquad (5.22)$$

For readers familiar with system theory, the characteristic polynomial is seen to be $\det (sI - A) = s^2 - (y_v + n_r)s + (y_v n_r + n_v U)$. For the missile to be asymptotically stable (called in missile literature *dynamically stable*) all the zeros must lie in the open left half s-plane, i.e. $-(y_v + n_r) > 0$ and $y_v n_r + n_v U > 0$. The first of these inequalities is always true because y_v and n_r are both negative. Because U is large the practical implication of the second inequality is that n_v must be greater than zero to give dynamic stability. This, of course is the same as the criterion for static stability mentioned previously. As an exercise the reader may care to prove that for a vehicle such as a torpedo where y_r and $y_v n_r$ are not negligible, the condition for dynamic stability is $1 - [n_v(y_r - U)/y_v n_r] > 0$, which is the so-called G-criterion.

5.5.2 Stability augmentation
If natural frequency or damping ratio of the missile characteristic polynomial are not satisfactory they may be changed by a combination of accelerometer and rate gyroscope feedback. Let the gyroscope feedback gain be k_r, and the accelerometer feedback gain be k_a, so that

$$\zeta = k_r r + k_a f_y + \zeta_c \qquad (5.23)$$

where ζ_c is the command from the missile guidance system. From eqn. (5.16), (assuming $\dot{p}$ and ϕ are zero)

$$f_y = \dot{v} + Ur + b_x \dot{r} \qquad (5.24)$$

Substituting for $\dot{v}$ and $\dot{r}$ from eqn. (5.22) gives

$$f_y = (y_v + b_x n_v)v + b_x n_r r + (y_\zeta + b_x n_\zeta)\zeta \qquad (5.25)$$

Equation (5.25) shows that if f_y is used to drive the rudder via eqn. (5.23), then measured rudder position scaled by $k_a(y_\zeta + b_x n_\zeta)$ is part of the drive signal. This type of feedback is known as an *algebraic loop* and if the gain is too high may lead to high-frequency stability problems. To minimize this the accelerometer is usually placed in front of the missile centre of gravity near the position

$$b_x = -y_\zeta/n_\zeta$$

Equation (5.25) then reduces to

$$f_y = (y_v - y_\zeta n_v/n_\zeta)v - (y_\zeta n_r/n_\zeta)r \qquad (5.26)$$

Substituting eqn. (5.26) and (5.23) into (5.22) gives

$$\begin{bmatrix} \dot{v} \\ \dot{r} \end{bmatrix} = \begin{bmatrix} y_v + k_1 y_\zeta & -U + k_2 y_\zeta \\ n_v + k_1 n_\zeta & n_r + k_2 n_\zeta \end{bmatrix} \begin{bmatrix} v \\ r \end{bmatrix} + \begin{bmatrix} y_\zeta \\ n_\zeta \end{bmatrix} \zeta \qquad (5.27)$$

where

$$k_1 = k_a(y_v - y_\zeta n_v/n_\zeta)$$

$$k_2 = k_r - k_a y_\zeta n_r/n_\zeta .$$

The matrix $K = [k_1 \ k_2]$ is called the *state feedback matrix*.

If the characteristic polynomial of state equation (5.27) is evaluated, and small terms ignored, the result is

$$s^2 - (y_v + n_r + y_\zeta k_1 + n_\zeta k_2)s + (n_v + n_\zeta k_1)U.$$

The constant term $(n_v + n_\zeta k_1)U$ of this polynomial is denoted by ω_n^2, where ω_n is the missile natural frequency. Since k_1 is purely a function of k_a, the missile natural frequency may be adjusted by accelerometer feedback. The coefficient of s is denoted by $2\mu\omega_n$, where μ is the missile damping ratio. For any given k_a (chosen to give the desired value of ω_n), the value of k_2 can be altered by changing the rate gyro feedback gain k_r to give the required missile damping.

The values of ω_n and μ so obtained are sometimes called the *controlled* but *unguided* natural frequency and damping ratio.

To obtain the missile characteristics including the guidance system, a model of the guidance geometry must be added.

5.5.3 The geometry of line-of-sight (LOS) guidance
The geometry of a missile following a line of sight (LOS) toward a target is shown on Fig. 5.5. The symbols shown on the diagram are

U velocity along the missile x-axis
v velocity along the missile y-axis
$\Delta\psi$ angle between the missile x-axis and the LOS
Δy length of perpendicular from LOS to missile

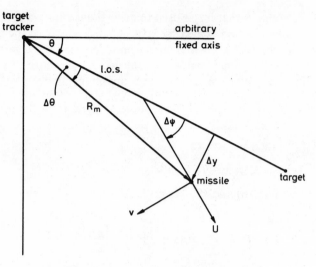

Fig. 5.5 *Geometry of line-of-sight guidance*

R_m range (distance) of missile from the ground-based tracking station
θ angle of LOS to an arbitrary fixed reference line through the tracker
$\Delta\theta$ angle between LOS and line joining tracker to missile.

From this figure,

$$\Delta\dot{y} = v \cos \Delta\psi + U \sin \Delta\psi - R_m\dot{\theta} \tag{5.28}$$

$$\Delta\psi = r - \dot{\theta} \tag{5.29}$$

Assuming that $\Delta\psi$ is small reduces eqn. (5.28) to

$$\Delta\dot{y} = v + U \Delta\psi - R_m\dot{\theta} \tag{5.30}$$

Equations (5.29) and (5.30) can be combined with eqn. (5.22) to give

$$\begin{bmatrix} \dot{v} \\ \dot{r} \\ \Delta\dot{\psi} \\ \Delta\dot{y} \end{bmatrix} = \begin{bmatrix} y_v & -U & 0 & 0 \\ n_v & n_r & 0 & 0 \\ 0 & 1 & 0 & 0 \\ 1 & 0 & U & 0 \end{bmatrix} \begin{bmatrix} v \\ r \\ \Delta\psi \\ \Delta y \end{bmatrix} + \begin{bmatrix} y_\zeta \\ n_\zeta \\ 0 \\ 0 \end{bmatrix} \zeta + \begin{bmatrix} 0 \\ 0 \\ -1 \\ -R_m \end{bmatrix} \theta \tag{5.31}$$

This is a typical 'aerospace model' in that it combines several rows of dynamics with several rows of geometry. The same pattern occurs in the aircraft equations derived in Section 5.6.

Consider the last row of eqn. (5.31)

$$\Delta\dot{y} = v + U\Delta\psi - R_m\dot{\theta}$$

Differentiating and substituting for $\Delta\dot{\psi}$ from row 3 gives

$$\Delta\ddot{y} = \dot{v} + Ur - (U + \dot{R}_m)\dot{\theta} - R_m\ddot{\theta}$$

If $\Delta\psi$ is small, then $U \doteq \dot{R}_m$. Hence

$$\Delta\ddot{y} = (\dot{v} + Ur) - (2\dot{R}_m\dot{\theta} + R_m\ddot{\theta}) \tag{5.32}$$

From eqn. (5.24), the term $\dot{v} + Ur$ is the lateral acceleration of the missile at its centre of gravity f_m, say. Elementary kinematics show that $R_m\ddot{\theta} + 2\dot{R}_m\dot{\theta}$ is the lateral acceleration of the LOS at a distance R_m from the tracker f_L, say. Thus

$$\Delta\ddot{y} = f_m - f_L \tag{5.33}$$

where

$$f_m = \dot{v} + Ur \tag{5.34}$$
$$f_L = R_m\ddot{\theta} + 2\dot{R}_m\dot{\theta} \tag{5.35}$$

Figure 5.6 shows a typical control structure arising from this model. An estimate of LOS acceleration is calculated on the ground from measured values of $\dot{\theta}$ and $\ddot{\theta}$, and this forms the basic guidance command to the missile. The displacement of the missile from the LOS is also measured and used as a 'double integral' feedback correction term. Compensation is necessary to give adequate stability to the geometrical (guidance) loop, and this may conflict with noise rejection requirements. Alternative feedback structures including optimal filtering, and full state feedback will probably occur to those readers who are control specialists.

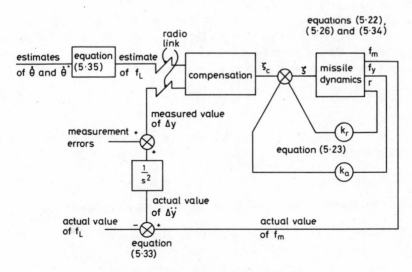

Fig. 5.6   *Missile LOS control block diagram*

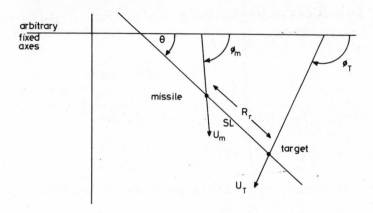

Fig. 5.7   *Homing geometry*

5.5.4 *Geometry of terminal homing*

Figure 5.7 shows the geometry of a missile homing onto a target. The sight line (SL) which joins the two differs from the LOS described in Section 5.5.3 because the 'end' is not fixed to the ground by a target tracker.

It is conventional when modelling the homing performance to replace the state variables v and ψ by

$$U_m = (U^2 + v^2)^{1/2}$$

$$\phi_m = \psi + \arctan \frac{v}{U}$$

In simulation studies this added complexity may be unnecessary. However, for analysis, if the missile dynamics are fast compared with the rates of change of geometry, then it may be assumed that $\dot{\phi}_m$ is controlled directly. The missile thus becomes equivalent to a particle with lateral acceleration

$$f_m = U_m \dot{\phi}_m \tag{5.36}$$

travelling at speed U_m.

Similarly the target has speed U_T directed at angle ϕ_T and has lateral acceleration f_T

$$f_T = U_T \dot{\phi}_T \tag{5.37}$$

The mathematical model is obtained by resolving along the SL:

$$\dot{R}_r = U_T \cos(\phi_T - \theta) - U_m \cos(\phi_m - \theta) \tag{5.38}$$

and perpendicular to the SL:

$$R_r \dot{\theta} = U_T \sin(\phi_T - \theta) - U_m \sin(\phi_m - \theta) \tag{5.39}$$

The way in which these equations are used depends upon the *homing policy*, i.e. the way in which the missile attacks the target. For example, *pursuit homing* has the policy

$$\phi_m = \theta \tag{5.40}$$

Thus the missile always points its velocity vector at the target.

Substituting eqns. (5.40) and (5.36) into (5.39) gives

$$f_m = \frac{U_m U_T}{R_r} \sin (\phi_T - \phi_m)$$

If the maximum achievable lateral acceleration of the missile is $|f_{max}|$, then to follow a pursuit trajectory

$$R_r \geq \frac{U_m U_T}{|f_{max}|} \sin (\phi_T - \phi_m) \tag{5.41}$$

This inequality is shown graphically on Fig. 5.8, and illustrates why missiles that employ this mode of guidance usually end in a tail chase (good if the missile has an infrared heat-seeking head).

One way of maintaining moderate acceleration levels is to use *proportional navigation* guidance

$$\dot{\phi}_m = k\theta \tag{5.42}$$

An interesting expression for the value of k can be obtained by linearizing about the trajectory which would give interception to a constant-velocity

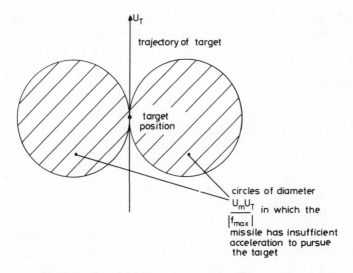

Fig. 5.8 *Pursuit homing constant acceleration contours*

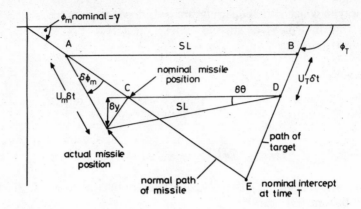

Fig. 5.9 *Proportional navigation homing geometry*

target. The similar triangles ABE and CDE of Fig. 5.9 show that if ϕ_T, U_T and U_m are constant, then ϕ_m should be constant. Further, the SL angle θ is constant. This is confirmed by eqn. (5.39). For $\dot{\theta} = 0$,

$$U_T \sin (\phi_T - \theta) - U_m \sin (\phi_m - \theta) = 0$$

Arbitrarily choose $\theta = 0$, and perturb the missile from the nominal trajectory $\phi_m = \gamma$ by $\delta\phi_m$. Let this perturbation cause deviations after time δt of δy and $\delta\theta$ as shown on Fig. 5.9.

Then from the figure,

$$\delta y = R_r \, \delta\theta \tag{5.43}$$

where R_r is the distance between missile and target. For small perturbations, the closing velocity of the missile and target is constant, U_r, say. Thus

$$U_r = \dot{R}_r \tag{5.44}$$

Differentiating eqn. (5.43) gives

$$\delta\dot{y} = R_r\dot{\theta} - U_r \, \delta\theta \tag{5.45}$$

Equation (5.39) for $\theta = 0$ is

$$R_r\dot{\theta} = U_T \sin \phi_T - U_m \sin (\gamma + \delta\phi_m)$$

$$= U_T \sin \phi_T - U_m \sin \gamma - U_m \, \delta\phi_m \cos \gamma$$

But γ has been chosen to make $U_T \sin \phi_T - U_m \sin \gamma$ zero. Thus the SL rotation caused by deviation $\delta\phi_m$ is

$$\dot{\theta} = - \frac{U_m \cos \gamma}{R_r} \, \delta\phi_m \tag{5.46}$$

Substitute eqn. (5.46) into (5.45) and differentiate

$$\delta\ddot{y} = -U_m \cos\gamma\ \dot{\phi}_m - U_r\dot{\theta}$$

Substituting again for $\dot{\theta}$ gives

$$\delta\ddot{y} = -U_m \cos\gamma\ (\dot{\phi}_m - \delta\phi_m R_r/U_r)$$

It is convenient to ignore the second term and assume that $\delta\ddot{y}$ can be regarded as an input to the model determined by the turn rate $\dot{\phi}_m$. Thus

$$\delta\ddot{y} = -U_m \cos\gamma\ \dot{\phi}_m \tag{5.47}$$

Now that δy, $\delta\dot{y}$ and $\delta\ddot{y}$ have been found in terms of the usual missile description (R_r, $\delta\theta$, U_m, etc.), the geometry can be expressed by the extremely simple model

$$\frac{d\delta y}{dt} = \delta\dot{y} \qquad \text{and} \qquad \frac{d\delta\dot{y}}{dt} = \delta\ddot{y}$$

In state form

$$\begin{bmatrix} \delta\dot{y} \\ \delta\ddot{y} \end{bmatrix} = \begin{bmatrix} 0 & 1 \\ 0 & 0 \end{bmatrix} \begin{bmatrix} \delta y \\ \delta\dot{y} \end{bmatrix} + \begin{bmatrix} 0 \\ 1 \end{bmatrix} \delta\ddot{y} \tag{5.48}$$

A technique which can now be applied is to find the state feedback matrix which minimizes the cost function

$$V = \delta y(T)^2 + \int_0^T \frac{\delta\ddot{y}^2}{\alpha^3}\ dt \tag{5.49}$$

The crucial term of eqn. (5.49) is $\delta y(T)$, the distance perpendicular to the *SL* by which the missile misses the target. However, the integral of $\delta\ddot{y}^2/\alpha^3$ puts a constraint on the missile's lateral acceleration whose magnitude is determined by an appropriate choice of α.

The method of minimizing functions such as V is well known to control engineers,[5] and basically consists of solving a non-linear matrix differential equation called the matrix Riccati equation. The particular form arising from (5.48) and (5.49) is unusual because it has an analytic solution, and results in the state feedback relationship

$$\delta\ddot{y} = -\frac{3\alpha^3}{3 + \alpha^3(R_r/U_r)^3} \left[\frac{R_r}{U_r}\ \left(\frac{R_r}{U_r}\right)^2 \right] \begin{bmatrix} \delta y \\ \delta\dot{y} \end{bmatrix} \tag{5.50}$$

Substituting for $\delta\ddot{y}$, $\delta\dot{y}$ and δy from eqns. (5.47), (5.43) and (5.45) gives

$$\dot{\phi}_m = \frac{3U_r\alpha^3(R_r/U_r)^3}{U_m \cos \gamma(3 + \alpha^3(R_r/U_r)^3)}\dot{\theta}$$

Clearly when the range is large compared with the closing speed

$$\alpha^3(R_r/U_r)^3 \gg 3 \tag{5.51}$$

(note the cube law) and

$$\dot{\phi}_m = \frac{3U_r}{U_m \cos \gamma}\dot{\theta} \tag{5.52}$$

Thus the value of k in the proportional navigation equation (5.42) is

$$\frac{3U_r}{U_m \cos \gamma}$$

If inequality (5.51) does not hold, the assumptions of the mathematical model are also suspect: for example $\delta\phi_m R_r/U_r$ should perhaps not be ignored in obtaining eqn. (5.47).

5.6 Model of the motion of an aeroplane

As a second example of the use of the equations developed in Section 5.4, consider the motion of a conventional aeroplane. The aerodynamic derivatives which are usually non-negligible are[2]

$$
\begin{array}{llllll}
X_u & X_w & X_\eta \\
Y_v & Y_r & Y_\zeta \\
Z_u & Z_w & Z_q & Z_{\dot{w}} & Z_{\dot{q}} & Z_\eta \\
L_v & L_p & L_r & L_{\dot{p}} & L_{\dot{r}} & L_\xi & L_\zeta \\
M_u & M_w & M_q & M_{\dot{w}} & M_{\dot{q}} & M_\eta \\
N_v & N_p & N_r & N_{\dot{p}} & N_{\dot{r}} & N_\xi & N_\zeta
\end{array}
$$

If the state vector is taken in the obvious order

$$[u \; v \; w \; p \; q \; r \; \phi \; \theta \; \psi \; h]^T,$$

an attractive pattern of zero and non-zero elements are seen when the equations are written in matrix form. A certain amount of trial and error on the mathematics, or alternatively and far better, an appreciation of the physical implications of the model, results in a block diagonal form if the state vector is reordered

$$[u\ w\ q\ \theta\ h : v\ p\ r\ \phi\ \psi]^T$$

$$
\begin{bmatrix}
m & 0 & 0 & 0 & 0 & & & & & \\
0 & m-Z_{\dot{w}} & -Z_{\dot{q}} & 0 & 0 & & & & & \\
0 & -M_{\dot{w}} & I_y-M_{\dot{q}} & 0 & 0 & & & 0 & & \\
0 & 0 & 0 & 1 & 0 & & & & & \\
0 & 0 & 0 & 0 & 1 & & & & & \\
& & & & & m & 0 & 0 & 0 & 0 \\
& & & & & 0 & I_x-L_{\dot{p}} & -L_{\dot{r}} & 0 & 0 \\
& & 0 & & & 0 & -N_{\dot{p}} & I_z-N_{\dot{r}} & 0 & 0 \\
& & & & & 0 & 0 & 0 & 1 & 0 \\
& & & & & 0 & 0 & 0 & 0 & 1
\end{bmatrix}
\begin{bmatrix}
\dot{u} \\ \dot{w} \\ \dot{q} \\ \dot{\theta} \\ \dot{h} \\ \dot{v} \\ \dot{p} \\ \dot{r} \\ \dot{\phi} \\ \dot{\psi}
\end{bmatrix}
$$

$$
=
\begin{bmatrix}
X_u & X_w & 0 & -mg & 0 & & & & & \\
Z_u & Z_w & Z_q+mU & 0 & 0 & & & & & \\
M_u & M_w & M_q & 0 & 0 & & & 0 & & \\
0 & 0 & 1 & 0 & 0 & & & & & \\
0 & -1 & 0 & U & 0 & & & & & \\
& & & & & Y_v & 0 & Y_r-mU & mg & 0 \\
& & & & & L_v & L_p & L_r & 0 & 0 \\
& & 0 & & & N_v & N_p & N_r & 0 & 0 \\
& & & & & 0 & 1 & 0 & 0 & 0 \\
& & & & & 0 & 0 & 1 & 0 & 0
\end{bmatrix}
\begin{bmatrix}
u \\ w \\ q \\ \theta \\ h \\ v \\ p \\ r \\ \phi \\ \psi
\end{bmatrix}
$$

$$
+
\begin{bmatrix}
X_\eta & 1 & & \\
Z_\eta & \epsilon & & \\
M_\eta & c_z & & 0 \\
0 & 0 & & \\
0 & 0 & & \\
& & 0 & Y_\zeta \\
& & L_\xi & L_\zeta \\
& 0 & N_\xi & N_\zeta \\
& & 0 & 0 \\
& & 0 & 0
\end{bmatrix}
\begin{bmatrix}
\eta \\ \partial E \\ \xi \\ \zeta
\end{bmatrix}
\tag{5.53}
$$

The equation comes from the following sources:

row 1	longitudinal force balance	1st eqn. of (5.15)
row 2	normal force balance	3rd eqn. of (5.15)
row 3	pitch moment balance	5th eqn. of (5.15)
row 4	pitch rate to pitch angle	2nd row of (5.12)
row 5	body velocity to reference velocity	eqn. (5.11)
row 6	lateral force balance	2nd eqn. of (5.15)

row 7   roll moment balance   4th eqn. of (5.15)
row 8   yaw moment balance   6th eqn. of (5.15)
row 9   roll rate to roll angle   1st row eqn. of (5.12)
row 10   yaw rate to yaw angle   3rd row eqn. of (5.12)

The gravitational forces and moments occurring in rows, 1, 2, 3, 6, 7 and 8 are specified by eqn. (5.17), the propulsion forces and moments by (5.18) and the aerodynamic forces and moments by the assumed non-zero derivatives listed above.

For physical reasons, which it is hoped are obvious to the reader, the first five rows are called the *longitudinal* equations of motion and the second five rows are called the *lateral* equations of motion. The block diagonal form shows that the motions are independent.

5.6.1 Longitudinal equations of motion

It can be seen from eqn. (5.53) that the longitudinal model is described by three dynamic equations and two geometrical equations. Together they give rise to three modes* of motion. Put the equations in to state form

$$\dot{\mathbf{x}} = A\mathbf{x} + B\mathbf{u} \tag{5.54}$$

by premultiplying by the inverse of the 'mass' matrix

$$\begin{bmatrix} m & 0 & 0 & 0 & 0 \\ 0 & m - Z_{\dot{w}} & -Z_{\dot{q}} & 0 & 0 \\ 0 & -M_{\dot{w}} & I_y - M_{\dot{q}} & 0 & 0 \\ 0 & 0 & 0 & 1 & 0 \\ 0 & 0 & 0 & 0 & 1 \end{bmatrix}$$

The inverse always exists because the Gershgorin discs do not allow any zero eigenvalues (i.e. $|m - Z_{\dot{w}}| > |Z_{\dot{q}}|$ and $|I_y - M_{\dot{q}}| > |M_{\dot{w}}|$). The longitudinal equation of motion becomes

$$\begin{bmatrix} \dot{u} \\ \dot{w} \\ \dot{q} \\ \dot{\theta} \\ \dot{h} \end{bmatrix} = \begin{bmatrix} x_u & x_w & 0 & -g & 0 \\ z_u & z_w & z_q + U & 0 & 0 \\ m_u & m_w & m_q & 0 & 0 \\ 0 & 0 & 1 & 0 & 0 \\ 0 & -1 & 0 & U & 0 \end{bmatrix} \begin{bmatrix} u \\ w \\ q \\ \theta \\ h \end{bmatrix} + \begin{bmatrix} x_\eta & 1 \\ z_\eta & z_e \\ m_\eta & m_e \\ 0 & 0 \\ 0 & 0 \end{bmatrix} \begin{bmatrix} \eta \\ e \end{bmatrix} \tag{5.55}$$

The lower-case letters for the derivatives indicate that to a good degree of accuracy (and in most cases exactly), $x_u = X_u/m$, $m_w = M_w/I_y$, etc. Note that the perturbation engine thrust ∂E has been replaced by the specific perturbation thrust $e = \partial E/m$. Naturally e is in units of acceleration, so element $(1, 2)$ of the input matrix turns out to be 1. If the characteristic polynomial

---

* Readers familiar with system theory will find that the name 'mode' is used in a similar but not quite identical sense to $\mu e^{\lambda t}$.

det $(sI - A)$ is evaluated for eqn. (5.55), it is usually found to be of the form $s(s^2 + 2\mu_q\omega_q s + \omega_q^2)(s^2 + 2\mu_p\omega_p s + \omega_p^2)$ corresponding to the *height integration* mode, *short period* (quick) mode and *phugoid* mode of motion, respectively.

The height integration mode. The last column of the A-matrix of eqn. (5.55) is zero. Thus h does not affect any other state variable, and the last row of the matrix therefore represents pure integration. This accounts for the s term in det $(sI - A)$. Clearly the last row and column may be removed without affecting the behaviour of the rest of the system, and eqn. (5.55) reduces to

$$\begin{bmatrix} \dot{u} \\ \dot{w} \\ \dot{q} \\ \dot{\theta} \end{bmatrix} = \begin{bmatrix} x_u & x_w & 0 & -g \\ z_u & z_w & z_q + U & 0 \\ m_u & m_w & m_q & 0 \\ 0 & 0 & 1 & 0 \end{bmatrix} \begin{bmatrix} u \\ w \\ q \\ \theta \end{bmatrix} + \begin{bmatrix} x_\eta & 1 \\ z_\eta & z_e \\ m_\eta & m_e \\ 0 & 0 \end{bmatrix} \begin{bmatrix} \eta \\ e \end{bmatrix} \qquad (5.56)$$

The short period mode. In a classical aircraft the forward speed changes very slowly compared with pitch motions, and it is reasonable to assume that these fast transients occur at effectively constant speed, say $u = 0$. (Recall that u is the perturbation in forward speed about the steady value U.)

The first row of (5.56) may therefore be neglected when investigating the fast pitch motion

$$\begin{bmatrix} \dot{w} \\ \dot{q} \\ \dot{\theta} \end{bmatrix} = \begin{bmatrix} z_w & z_q + U & 0 \\ m_w & m_q & 0 \\ 0 & 1 & 0 \end{bmatrix} \begin{bmatrix} u \\ w \\ \theta \end{bmatrix} + \begin{bmatrix} z_\eta & z_e \\ m_\eta & m_e \\ 0 & 0 \end{bmatrix} \begin{bmatrix} \eta \\ e \end{bmatrix} \qquad (5.57)$$

The pitch angle now has no effect on the motion because the last column of eqn. (5.57) is zero. The equations describing the fast pitch transients therefore reduce to

$$\begin{bmatrix} \dot{w} \\ \dot{q} \end{bmatrix} = \begin{bmatrix} z_w & z_q + U \\ m_w & m_q \end{bmatrix} \begin{bmatrix} w \\ q \end{bmatrix} + \begin{bmatrix} z_\eta & z_e \\ m_\eta & m_e \end{bmatrix} \begin{bmatrix} \eta \\ e \end{bmatrix} \qquad (5.58)$$

It is usually found on a classical aircraft that $|z_q| \ll U$. Also the engine thrust is almost along the Ox axis making z_e and m_e small. Equation (5.58) further reduces to

$$\begin{bmatrix} \dot{w} \\ \dot{q} \end{bmatrix} = \begin{bmatrix} z_w & U \\ m_w & m_q \end{bmatrix} \begin{bmatrix} w \\ q \end{bmatrix} + \begin{bmatrix} z_\eta \\ m_\eta \end{bmatrix} \eta \qquad (5.59)$$

The characteristic polynomial det $(sI - A)$ of this state equation is

$$s^2 + (-z_w - m_q)s + (z_w m_q - m_w U) \qquad (5.60)$$

It is usually found that this is a very good approximation of the factor $(s^2 + 2\mu_q\omega_q s + \omega_q^2)$ appearing in the full characteristic polynomial of (5.55).

Readers may like to compare eqn. (5.59) with eqn. (5.22) which described the lateral behaviour of a guided missile. Allowing for the sign change in U

which is due to v and r having a different mutual geometrical relationship to that of w and q (corresponding sign changes occur in the aerodynamic derivatives), the equations are identical and describe essentially the same behaviour. Using the same arguments as the missile case, the practical implication of polynomial (5.60) is that the short-period mode is stable if $m_w < 0$.

Readers should study Fig. 5.1 to convince themselves that $m_w < 0$ and $n_v > 0$ both imply weathercock stability (a vehicle suspended at its c.g. will 'turn into wind') and it is merely geometry which causes the sign reversal.

The Phugoid mode. Because the short-period mode is fast, it may be assumed that its differential equations are instantaneously satisfied compared with the slower motions. The zero-input ($\eta = e = 0$) version of eqn. (5.56) therefore becomes

$$\begin{bmatrix} \dot{u} \\ 0 \\ 0 \\ \dot{\theta} \end{bmatrix} = \begin{bmatrix} x_u & x_w & 0 & -g \\ z_u & z_w & U & 0 \\ m_u & m_w & m_q & 0 \\ 0 & 0 & 1 & 0 \end{bmatrix} \begin{bmatrix} u \\ w \\ q \\ \theta \end{bmatrix} \tag{5.61}$$

Solving the algebraic second and third rows to obtain w and q in terms of u, and substituting into the first and fourth rows gives

$$\begin{bmatrix} \dot{u} \\ \\ \dot{\theta} \end{bmatrix} = \begin{bmatrix} x_u - x_w \dfrac{m_q z_u - m_u U}{m_q z_w - m_w U} & -g \\ \\ -\dfrac{z_u m_w - m_u z_w}{U m_w - m_q z_w} & 0 \end{bmatrix} \begin{bmatrix} u \\ \\ \theta \end{bmatrix} \tag{5.62}$$

The characteristic polynomial is

$$s^2 + \left[x_u - x_w \left(\frac{m_q z_u - m_u U}{m_q z_w - m_w U} \right) \right] s - g \left(\frac{z_u m_w - m_u z_w}{U m_w - m_q z_w} \right)$$

It is usually found that this is a reasonable approximation of the factor $(s^2 + 2\mu_p \omega_p s + \omega_p^2)$ appearing in the full characteristic polynomial of (5.55).

Physical interpretation of the modes. The interpretation of height integration is rather obvious; if the aircraft is climbing there is no feedback to prevent it, so height continues to increase.

The short-period motion is due to the 'arrow stability' of the aircraft: the distance of the aerodynamic centre from the c.g. and the magnitude of the various aerodynamic damping terms.

The phugoid mode has a more subtle but rather interesting interpretation. Assume (and this is true for a typical subsonic classical aircraft) that $|z_u m_w| \gg |m_u z_w|$ and $|U m_w| \gg |m_q z_w|$. Thus $\omega_p \doteq \sqrt{-g z_u / U}$. The negative sign causes no concern because z_u is negative. Now under equilibrium conditions the total lift is proportional to U^2, and in the opposite direction of Oz.

$$Z = -kU^2 = -mg \tag{5.63}$$

and z_u can be calculated from

$$z_u = \frac{1}{m}\frac{\partial Z}{\partial u}$$

Now $\partial u = \partial U$, so

$$z_u = \frac{1}{m}\frac{\partial}{\partial U}(-kU^2) = -\frac{2kU}{m}$$

Substituting from (5.63) for k gives

$$z_u = -\frac{2g}{U}$$

which can be substituted into the simplified expression for ω_p to give

$$\omega_p \doteqdot \sqrt{2}\,\frac{g}{U} \tag{5.64}$$

Now consider a very simple model of a constant energy vehicle

$$\tfrac{1}{2}mU^2 + mgH = \text{constant}$$

supported by a force proportional to speed squared

$$m\frac{d^2H}{dt^2} = kU^2 - mg$$

Perturbations about these two equations give

$$Uu + gh = 0 \quad\text{and}\quad m\frac{d^2h}{dt^2} = 2kUu$$

Eliminating u between them, and substituting for k from eqn. (5.63) gives

$$\ddot{h} = -2\left(\frac{g}{U}\right)^2 h$$

Thus this very simple model gives an undamped height oscillation of the same frequency as eqn. (5.64). This analysis therefore indicates that the phugoid is basically an oscillation reflecting an interchange of potential and kinetic energy, although this is heavily disguised in eqn. (5.62).

5.6.2 *Lateral equations of motion*

The last five rows of eqn. (5.53) describe the lateral motion of a classical aircraft. They can be put into state form by a similar procedure to that employed in Section 5.6.1.

$$
\begin{bmatrix} \dot{v} \\ \dot{p} \\ \dot{r} \\ \dot{\phi} \\ \dot{\psi} \end{bmatrix} = \begin{bmatrix} y_v & 0 & y_r - U & g & 0 \\ l_v & l_p & l_r & 0 & 0 \\ n_v & n_p & n_r & 0 & 0 \\ 0 & 1 & 0 & 0 & 0 \\ 0 & 0 & 1 & 0 & 0 \end{bmatrix} \begin{bmatrix} v \\ p \\ r \\ \phi \\ \psi \end{bmatrix} + \begin{bmatrix} 0 & y_\zeta \\ l_\xi & l_\zeta \\ n_\xi & n_\zeta \\ 0 & 0 \\ 0 & 0 \end{bmatrix} \begin{bmatrix} \xi \\ \zeta \end{bmatrix}
\tag{5.65}
$$

Like the longitudinal case, the first three rows are dynamics and the last two geometry. The lower-case letters indicate multiplication by the inverse of the 'mass' matrix so that to a high degree of accuracy $y_v = Y_v/m$, $l_v = L_v/I_x$, etc. If the characteristic polynomial det $(sI - A)$ is evaluated it will usually be of the form

$$
s\left(s + \frac{1}{\tau_r}\right)(s^2 + 2\mu_d \omega_d s + \omega_d^2)\left(s + \frac{1}{\tau_s}\right)
\tag{5.66}
$$

which correspond to the *heading integration* mode, the *roll subsidence* mode, the *dutch roll* and the *spiral* mode, respectively.

Heading integration mode. Like the height integration mode, this arises because the last column of the state matrix is zero. Equation (5.65) can therefore be reduced to

$$
\begin{bmatrix} \dot{v} \\ \dot{p} \\ \dot{r} \\ \dot{\phi} \end{bmatrix} = \begin{bmatrix} y_v & 0 & y_r - U & g \\ l_v & l_p & l_r & 0 \\ n_v & n_p & n_r & 0 \\ 0 & 1 & 0 & 0 \end{bmatrix} \begin{bmatrix} v \\ p \\ r \\ \phi \end{bmatrix} + \begin{bmatrix} 0 & y_\zeta \\ l_\xi & l_\zeta \\ n_\xi & n_\zeta \\ 0 & 0 \end{bmatrix} \begin{bmatrix} \xi \\ \zeta \end{bmatrix}
\tag{5.67}
$$

Roll subsidence mode. For the model of a classical aircraft, τ_r in polynomial (5.66) is small compared with τ_s or with $1/\omega_d$. It is caused primarily by the high damping of the wings in roll, which can be described by

$$
\dot{p} = l_p p
$$

Usually the approximation $\tau_r = -1/l_p$ is quite accurate.

Dutch roll mode. Because the roll subsidence decays so rapidly, the zero-input version of eqn. (5.67) can be written

$$
\begin{bmatrix} \dot{v} \\ 0 \\ \dot{r} \\ \dot{\phi} \end{bmatrix} = \begin{bmatrix} y_v & 0 & y_r - U & g \\ l_v & l_p & l_r & 0 \\ n_v & n_p & n_r & 0 \\ 0 & 1 & 0 & 0 \end{bmatrix} \begin{bmatrix} v \\ p \\ r \\ \phi \end{bmatrix}
$$

The second row can be solved to give

$$p = -\frac{1}{l_p}(l_v v + l_r r) \tag{5.68}$$

Ignoring the effect of gravitation ϕ has no effect on the motion, so that the last row may be discarded. Assuming that $|y_r| \ll U$ the equation reduces to

$$\begin{bmatrix} \dot{v} \\ \dot{r} \end{bmatrix} = \begin{bmatrix} y_v & -U \\ n_v - \dfrac{n_p l_v}{l_p} & n_r - \dfrac{n_p l_r}{l_p} \end{bmatrix} \begin{bmatrix} v \\ r \end{bmatrix} \tag{5.69}$$

The characteristic polynomial det $(sI - A)$ is

$$s^2 + \left(-y_v - n_r + \frac{n_p l_r}{l_p}\right)s + \left(y_v n_r + n_v U - \frac{y_v n_p l_r}{l_p} - \frac{n_p l_v U}{l_p}\right)$$

This polynomial is usually a close approximation to the factor $(s^2 + 2\mu_d \omega_d s + \omega_d^2)$ in the characteristic polynomial of eqn. (5.65).

Comparing eqn. (5.69) with (5.22) and (5.59), it is clear that the Dutch roll has similar origins to the yawing motion of the missile, or the short-period pitch oscillations of the aircraft. However, it is modified by the accompanying rolling motion which is necessary to satisfy eqn. (5.68).

Spiral mode. The final time constant in the characteristic polynomial τ_s is usually very slow indeed.

It can be approximated by solving the first three rows of the equation

$$\begin{bmatrix} 0 \\ 0 \\ 0 \\ \dot{\phi} \end{bmatrix} = \begin{bmatrix} y_v & 0 & -U & g \\ l_v & l_p & l_r & 0 \\ n_v & n_p & n_r & 0 \\ 0 & 1 & 0 & 0 \end{bmatrix} \begin{bmatrix} v \\ p \\ r \\ \phi \end{bmatrix}$$

to eliminate v and r. Putting $p = \dot{\phi}$ (last row) this gives

$$\dot{\phi} = g\left[\frac{y_v(n_r l_p - l_r n_p)}{n_r l_v - l_r n_v} - \frac{U(n_v l_p - l_v n_p)}{n_v l_r - l_v n_r}\right]^{-1}\phi$$

Usually the first term in the square brackets is small compared with the second, and a reasonable approximation to τ_s is

$$\tau_s = \frac{U}{g}\left(\frac{n_v l_p - l_v n_p}{n_v l_r - l_v n_r}\right) \tag{5.70}$$

Physical interpretation of the modes. The heading integration mode, like the height integration mode, is purely geometrical in nature and reflects the fact that there is no feedback of heading angle into the equations of motion.

The roll subsidence mode is caused by the high roll damping effect of the wing.

The Dutch roll is basically the yaw equivalent to the pitch short-period oscillation. However, the motion couples into roll, and this modifies the characteristic polynomial. In some aircraft the effect of the roll coupling on natural frequency and damping ratio is quite small. i.e.

$$\left| \frac{n_p l_v}{l_p} \right| \ll |n_v| \quad \text{and} \quad \left| \frac{n_p l_r}{l_p} \right| \ll |n_r|.$$

However, the impression gained by an observer in the aircraft may be of a predominantly rolling oscillation because of the large values of p caused via eqn. (5.68).

The spiral mode is a gravitional effect. When the aircraft rolls, gravity causes it to sideslip and yaw. The resulting forces and moments may cause ϕ to decrease giving a stable spiral mode, but on many aircraft the resulting motion is unstable. However, it is so slow that only in exceptional circumstances do pilots have difficulty in controlling it.

5.6.3 Aircraft control-system design

Although this chapter is concerned with modelling rather than control, a few remarks about the use of the model obtained in Sections 5.6.2 and 5.6.3 are in order. Traditionally the control system has been designed in two stages: the inner loop and the outer loop. The function of the inner loop is to make the aircraft easy and pleasant to fly, and is often called a *stability augmentation system* (SAS). The outer loop is to replace the pilot for certain simple flight manoeuvres such as maintaining height, maintaining speed, turning onto a specified heading, climbing at a specified rate etc. This system is often called an *autopilot*. Outside both of these loops is the *navigation loop*. Commands derived from the inertial navigation system or radio navigation aids can be fed into the autopilot to demand heading changes. A block diagram of these functions is shown on Fig. 5.10.

In more recent aircraft there may be additional functions such as structural mode control, gust alleviation, automatic landing, remote weapon control, etc.

All of these areas provide interesting and challenging design problems, not least in providing hardware and software of sufficiently high integrity. Current discussion centres around the requirement that the probability of a catastrophic accident due to control system failure should be less than 10^{-9} per hour for passenger aircraft. (This is about ten thousand times better than the probability of being killed or seriously injured in a car accident in Britain). Some idea of the difficulty in proving that a design meets this specification may be gained by equating 10^9 hours with one hundred thousand years.

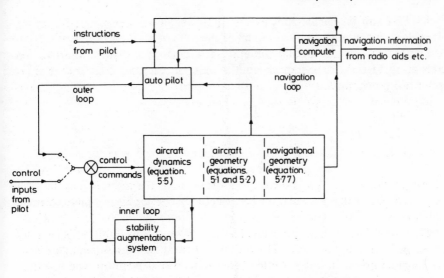

Fig. 5.10 *Block diagram of traditional aircraft control loop*

To return to the more traditional aspects of control system design, the SAS is a particularly interesting case because of the difficulty in specifying whether an aircraft handles well or badly. Reference 5 must be one of the most comprehensive and detailed specifications of a particular classs of control system performance that has ever been written, and is applied very widely as a design criterion. For aircraft whose dynamics are rather different to those of a classical aircraft, a very simple idea has been used to assess longitudinal handling, called the C* criterion[6].

It assumes that at low speeds a pilot senses longitudinal response visually via the pitch rate q. At high speeds he senses it by acceleration f_z through the 'seat of his pants'. A speed, U_{co} is chosen at which the two effects are assumed to be sensed equally by the average pilot. From eqn. (5.16), under steady pitch rate conditions $f_z = -Uq$. Therefore the expression $f_z + U_{co}q$ weights the terms equally at $U = U_{co}$. It gives more weight to acceleration when $U > U_{co}$, and more weight to pitch rate when $U < U_{co}$. The parameter C^* is defined by (the bar indicates Laplace transformation)

$$\bar{C}^* = \frac{1}{g}\left(\frac{\bar{f}_z}{\bar{\eta}} + U_{co}\frac{\bar{q}}{\bar{\eta}}\right)$$

and either the Bode plot of $\bar{C}^*(j\omega)$ or the step response of $C^*(t)$ must lie within certain regions. On some aircraft the control laws are continuously modified in flight by means of a so-called C^* box to keep the criterion satisfied.

5.6.4 Model of the human pilot

It is clear that a mathematical model of the human pilot is complementary to the specifications of handling quality. There are many variations possible according to the circumstances,[7] but a widely accepted model is that of the pilot being represented by the transfer function

$$P(s) = K \frac{e^{-\tau_d s}}{\tau_e s + 1} \cdot \frac{\alpha \tau_a s + 1}{\tau_a s + 1}$$

The values of time delay τ_d and the time constant τ_e depend upon the pilot's fatigue level, workload, physical health etc., and both lie between 1/20 and 1/4 seconds.

The parameters K, τ_a and α are instinctively chosen by the pilot to make the transfer function $Q(s)$ shown on Fig. 5.11 approximate to $\omega_c e^{-\tau_d s}/s$. Ignoring the pure time delay τ_d, this implies a closed loop behaviour of $\omega_c/(s + \omega_c)$, so ω_c can be considered as the bandwidth which the pilot considers necessary for adequate control. The pilot task is difficult if the bandwidth requirement is too high ($1/\omega_c$ fast compared with τ_e and τ_d), or the behaviour $\omega_c e^{-\tau_d s}/s$ cannot be achieved, or can only be achieved with excessive phase advance (a high value of α). For example the manual control of a double integrator plant $1/s^2$ is difficult because the desired $P(s)$ is $\omega_c e^{-\tau_d s}.s$ showing that τ_e must be small compared with $1/\omega_c$ and the phase advance required is 90°.

One disadvantage of the usual models of pilot behaviour is that they may be difficult to apply in the design situation. A very simple criterion was developed in Reference 8. It is assumed that the pilot has transfer function $K/0·25s + 1$. The designer attempts to choose a value of K which gives the dominant system poles a damping ratio of 0·7. If this is possible, and other factors (such as the stability of the remaining poles, system bandwidth, steady error accuracy etc.) are within specification, then the pilot will rate the handling as 'easy'.

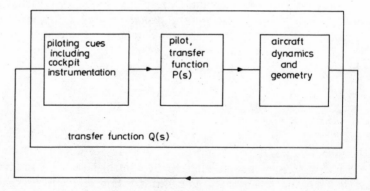

Fig. 5.11   *Human Pilot Block Diagram*

5.7 Model of an inertial navigation unit

A mathematical model of a gimballed inertial navigation unit (INU) will serve as a third example of the use of the equations developed in earlier sections. Traditionally the behaviour of INU's have been described in terms of the equations of vector mechanics. Rather than rewrite Section 5.3 in this language, the matrix description is retained. For the purposes of this chapter this has the advantage of continuity of style, and there is a further advantage which is mentioned in Section 5.7.3. The disadvantage is that the standard texts on IN may look unfamiliar.

The principle of operation of an INU may be understood from Fig. 5.12(*a*), which shows an accelerometer mounted on a railway truck. The displacement

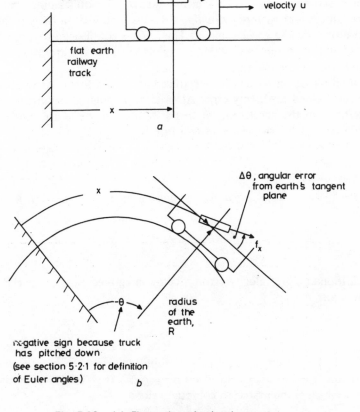

Fig. 5.12 (a) *Flat earth navigational system*
(b) *Round earth navigational system*

of the truck along the track could perhaps be deduced from double integration of the accelerometer output. The state equation is

$$\begin{bmatrix} \dot{x} \\ \dot{u} \end{bmatrix} = \begin{bmatrix} 0 & 1 \\ 0 & 0 \end{bmatrix} \begin{bmatrix} x \\ u \end{bmatrix} + \begin{bmatrix} 0 \\ 1 \end{bmatrix} f_x$$

Clearly the minimum polynomial of the state matrix is s^2. There is thus a repeated imaginary zero and the measurement of x is unstable. It is remarkable that the equivalent procedure for the vastly more complicated situation of a spherical rotating earth can give stable (although not asymptotically stable) navigation. The trick is to pivot the accelerometer on gimbals so that gravitation can stabilize the accelerometer errors.

To illustrate this consider the 'round stationary earth' railway track shown in Fig. 5.12(b). The distance x is now given by

$$x = -R\theta$$

so that distance may be determined by the angle through which the accelerometer axis has rotated. To control angular motion assume that a device called an *integrating rate gyroscope* (IRG) is mounted on the gimballed platform alongside the accelerometer. The IRG is coupled up to a motor on the gimbal system in such a way that a command to the IRG rotates the platform at any desired angular rate $\dot{\theta}$.

Suppose that the system has produced an accelerometer error Δf_x which in turn has caused a velocity error ΔU and an angular misalignment error $\Delta \theta$. Now to keep the accelerometer in the tangent plane of the earth, the IRG should be rotating the platform at a rate

$$\dot{\theta} = -\frac{U}{R} \tag{5.71}$$

However, the velocity error ΔU is causing rate of change of misalignment error

$$\Delta \dot{\theta} = -\frac{\Delta U}{R} \tag{5.72}$$

Additionally, the velocity error is being integrated to produce the displacement error.

$$\Delta \dot{x} = \Delta U \tag{5.73}$$

Because the accelerometer is tilted at $\Delta \theta$ to the tangent plane of the earth it senses not only the true acceleration $\dot{U}$ and its error Δf_x, but also a component of the gravitational field, $g \sin \Delta \theta \doteq g \, \Delta \theta$. It is this total accelerometer output which is integrated to compute velocity

$$U + \Delta U = \int (\dot{U} + \Delta f_x + g \, \Delta \theta) \, dt$$

Differentiating both sides gives

$$\Delta \dot{U} = \Delta f_x + g \, \Delta \theta \tag{5.74}$$

Writing eqns. (5.72), (5.73) and (5.74) in state form gives

$$\begin{bmatrix} \Delta \dot{x} \\ \Delta \dot{U} \\ \Delta \dot{\theta} \end{bmatrix} = \begin{bmatrix} 0 & 1 & 0 \\ 0 & 0 & g \\ 0 & -\dfrac{1}{R} & 0 \end{bmatrix} \begin{bmatrix} \Delta x \\ \Delta u \\ \Delta \theta \end{bmatrix} + \begin{bmatrix} 0 \\ 1 \\ 0 \end{bmatrix} \Delta f_x \tag{5.75}$$

The minimum polynomial of the state matrix is

$$s\left(s^2 + \frac{g}{R}\right) \tag{5.76}$$

which implies stability, although not asymptotic stability. Notice that the limit cycle has a frequency $\sqrt{g/R}$. The period of this is about 84 minutes and is called the Schuler period. The transfer function matrix, $(sI - A)^{-1}B$ is given by

$$\begin{bmatrix} \overline{\Delta x} \\ \overline{\Delta U} \\ \overline{\Delta \theta} \end{bmatrix} = \begin{bmatrix} \dfrac{1}{s^2 + g/R} \\ \dfrac{s}{s^2 + g/R} \\ \dfrac{1}{R(s^2 + g/R)} \end{bmatrix} \overline{\Delta f_x}$$

This shows that the zero of polynomial (5.76) at $s = 0$ is uncontrollable with respect to Δf_x, and hence the error Δx caused by a step in Δf_x is bounded and purely oscillatory with a period of 84 minutes.

The interested reader may like to anticipate a result presented in Section 5.7.3 and show that this happy circumstance does not occur if the error lies in the IRG instead of the accelerometer. [Replace eqn. (5.72) by

$$\Delta \dot{\theta} + \dot{\theta}_e = -\frac{\Delta U}{R}$$

where $\dot{\theta}_e$ is the IRG error.]

5.7.1 Spherical earth geometry

In Section 2.1, reference axes $O_0 x_0$, $O_0 y_0$, $O_0 z_0$ were defined in the tangent plane of the earth. It is now convenient to regard then as attached to the platform which, of course, should support accelerometers in this plane. In the past the orientation of $O_0 x_0 y_0 z_0$ adopted in various publications has been rather arbitrary, but for the reasons given at the end of Section 5.7.4 the orientation of $O_0 x_0$ east, $O_0 y_0$ south and $O_0 z_0$ down (ESD axes) are chosen for this chapter.

Let the platform have velocities V_x, V_y, V_z along these axes. Then the simple geometry of Fig. 5.13 shows that

$$\dot{\lambda} = -V_y/(R + h)$$
$$\dot{\Lambda} = [(V_x/(R + h)]\sec\lambda \qquad (5.77)$$
$$\dot{h} = -V_z$$

where λ is latitude, Λ is longitude and h is the height of O_0 above the surface of the earth.

Note that V_x, V_y, V_z are velocities with respect to a tangent plane rotating with the earth, and at a distance $(R + h)$ from the centre of the earth. If $V_x = V_y = V_z = 0$, the platform would remain at a height h above a particular geographical point on the earth's surface.

Equations (5.77) represent the transformation of V_x, V_y, V_z into 'conventional' navigation co-ordinates. For special purposes other transformations may be used—for example grid co-ordinates of a given map projection.

5.7.2 Platform gimbal geometry and precession rates
Let the aircraft be at rest on the surface of the earth, and $Oxyz$ lined up with $O_0 x_0 y_0 z_0$ (the ESD axes). Then a brief look at the definitions of Euler angles given in Section 5.2.1 will convince the reader that during the subsequent flight the gimbal system will measure Φ, Θ and Ψ directly, if the outer pivot is aligned along $O_0 x_0$, the intermediate pivot is aligned along $O_0 y_0$ and the inner pivot is aligned along $O_0 z_0$. Now the purpose of the gimbal system is to keep the platform isolated from the angular manoeuvres performed by the

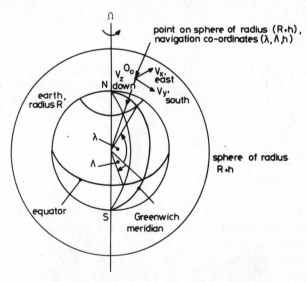

Fig. 5.13   *Spherical earth navigation geometry*

aircraft. The necessary pivot velocities are given by eqn. (5.2). It can be seen that $\dot{\Phi}$ and $\dot{\Psi} \to \infty$ as $\Theta \to \pm 90°$. This situation is known as *gimbal lock*, and reflects the mechanical situation of two pivots (outer and inner) becoming aligned onto a common axis, and one degree of kinematic freedom being consequently lost. To overcome this a fourth redundant gimbal is often slaved to the motion of the other three so that freedom is maintained at all orientations.

Let the rotation rates necessary to keep the platform aligned with ESD axes be ω_x, ω_y and ω_z about $O_0 x_0$, $O_0 y_0$ and $O_0 z_0$, respectively. These are called the *precession rates* of the platform. Each is made up of two components: (i) compensation for the rotation of the earth and (ii) compensation for the movement of the platform over the earth [a more complicated version of eqn. (5.71)]. Simple geometry applied to Fig. (5.13) gives

$$\omega_x = 0 + V_y/(R + h)$$
$$\omega_y = -\Omega \cos \lambda - V_x/(R + h) \qquad (5.78)$$
$$\omega_z = -\Omega \sin \lambda - [V_x/(R + h)] \tan \lambda$$

The first term in each equation compensates for the earth's rotation, and the second for the platform's geographical velocity.

The angular velocities at the pivots to give the required values of ω_x, ω_y and ω_z can be obtained from a transformation similar (but not identical) to eqn. (5.2). A fourth gimbal makes the equations more complicated and introduces a constraint relationship. It can be seen that the value of ω_z may be large at high latitudes. For transpolar navigation Ox_0 and Oy_0 are allowed to wander from east and south, their orientation being stored in a computer. This complicates the navigation equations by an extra angular resolution, and will not be considered in this chapter.

Calculation of platform velocity. The velocities V_x, V_y and V_z are basically deduced from the three accelerometer outputs a_x, a_y and a_z. Equation (5.7) of Section 5.3.2 holds, but the variables are now in terms of platform axes, not aircraft axes. Thus

$$\begin{bmatrix} f_x \\ f_y \\ f_z \end{bmatrix} \to \begin{bmatrix} a_x \\ a_y \\ a_z \end{bmatrix} \qquad (5.79)$$

$$\begin{bmatrix} P \\ Q \\ R \end{bmatrix} \to \begin{bmatrix} \omega_x \\ \omega_y \\ \omega_z \end{bmatrix} \qquad (5.80)$$

$$\begin{bmatrix} U \\ V \\ W \end{bmatrix} \to \begin{bmatrix} V_x \\ V_y \\ V_z \end{bmatrix} + \begin{bmatrix} V_{xe} \\ V_{ye} \\ V_{ze} \end{bmatrix} \qquad (5.81)$$

where V_{xe}, V_{ye} and V_{ze} are due to the earth's movement. For aircraft navigation the rotation of the earth around the sun is usually ignored, so

$$\begin{bmatrix} V_{xe} \\ V_{ye} \\ V_{ze} \end{bmatrix} = \begin{bmatrix} \Omega(R+h)\cos\lambda \\ 0 \\ 0 \end{bmatrix} \tag{5.82}$$

Differentiating (5.81) and (5.82) and substituting (5.77) gives

$$\begin{bmatrix} \dot{U} \\ \dot{V} \\ \dot{W} \end{bmatrix} \rightarrow \begin{bmatrix} \dot{V}_x + V_y\Omega\sin\lambda - V_z\Omega\cos\lambda \\ \dot{V}_y \\ \dot{V}_z \end{bmatrix} \tag{5.83}$$

Let the components of mass attraction along $O_0 x_0$, $O_0 y_0$ and $O_0 z_0$ be g_x, g_y and g_z, respectively. Equation (5.7) may now be written

$$\dot{V}_x = -2V_y\Omega\sin\lambda + 2V_z\Omega\cos\lambda + \frac{1}{R+h}$$

$$\times (V_z V_x - V_y V_x\tan\lambda) + [g_x] + a_x$$

$$\dot{V}_y = -2V_x\Omega\sin\lambda + \frac{1}{R+h}(V_x^2\tan\lambda + V_y V_z)$$

$$+ [\Omega^2(R+h)\cos\lambda\sin\lambda + g_y] + a_y \tag{5.84}$$

$$\dot{V}_z = -2V_x\Omega\cos\lambda - \frac{1}{R+h}(V_x^2 + V_y^2)$$

$$+ [-\Omega^2(R+h)\cos^2\lambda + g_z] + a_z$$

The terms in square brackets are due to the mass attraction and rotation of the earth, and depend upon λ, Λ and h. Their values are stored in the computer, and the raw acceleration signal compensated to give

$$a_x' = g_x + a_x$$
$$a_y' = \Omega^2(R+h)\cos\lambda\sin\lambda + g_y + a_y \tag{5.85}$$
$$a_z' = -\Omega^2(R+h)\cos^2\lambda + g_z + a_z$$

A block diagram for the solution of the navigation equations is shown in Fig. (5.14).

5.7.3 Accuracy of the inertial navigation unit

To assess the accuracy of the INU, a perturbation model of eqns. (5.77), (5.78), (5.84) and (5.85) are formed. For example the first equation of (5.78) becomes

$$\Delta\omega_x = \frac{\Delta V_y}{R+h} - \frac{V_y\Delta h}{(R+h)^2} \tag{5.86}$$

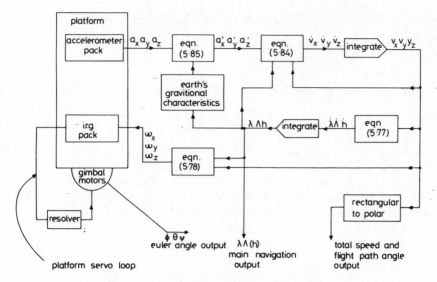

Fig. 5.14   *Block Diagram of Inertial Navigation Unit*

To these equations are added all the identifiable error sources. The resulting model then predicts the effect that each error has on navigational accuracy. Clearly the equations are rather complicated, although straightforward to
produce, and there is little point in quoting them here. However, to illustrate
the incredible degree of precision required for satisfactory performance, consider the effect of a drift error ω_{xe} in the $O_0 x_0$-axis IRG. Let the INU be
geographically stationary $(V_x = V_y = V_z = 0)$ and at sea level $(h = 0)$.

Assume that the drift error causes a velocity error in the $O_0 y_0$ direction of
ΔV_y. Then from eqn. (5.86)

$$\Delta \dot{\omega}_x = \omega_{xe} + \frac{\Delta V_y}{R}$$

This false rotation rate causes a misalignment error $\Delta \beta$ to develop between
the tangent plane and $O_0 y_0$

$$\Delta \dot{\beta} = \Delta \omega_x$$

A component of the gravitational field is thus sensed by the $O_0 y_0$ accelerometer, so that a_y' is not zero, but equal to $-g \, \Delta \beta$. Linearization of the
second equation of (5.84) is therefore $\Delta \dot{V}_y = -g \, \Delta \beta$.

The error state equation is therefore

$$\begin{bmatrix} \Delta \dot{\beta} \\ \Delta \dot{V}_y \end{bmatrix} = \begin{bmatrix} 0 & \dfrac{1}{R} \\ -g & 0 \end{bmatrix} \begin{bmatrix} \Delta \beta \\ \Delta V_y \end{bmatrix} + \begin{bmatrix} 1 \\ 0 \end{bmatrix} \omega_{xe} \qquad (5.87)$$

Solution of this for a constant value of drift rate $\omega_{xe} = \Omega_{xe}$ gives

$$\Delta V_y = R\left(\cos\sqrt{\frac{g}{R}}t - 1\right)\Omega_{xe}$$

This is an oscillatory error at the Schuler period with average value

$$\Delta V_y = -R\Omega_{xe}$$

This equation shows that a navigational accuracy of 1 nautical mile per hour, implies an IRG drift rate of one minute of arc per hour: $0\cdot017°/h$ or one revolution in two-and-a-half years. In fact, to allow for other INU errors, the IRG's are considerably better than this, and it is remarkable that instruments with this level of performance can be made at all, let alone sufficiently light and robust for the aircraft environment.

5.7.4 Strapdown inertial navigation units

Over the last decade there has been increasing research into INU's which replace the mechanical gimbal system of the platform with an appropriate computer algorithm. The gyroscopes and accelerometers are now fixed directly to the airframe, and the navigator is called a *strapdown* system. A major difficulty has been the design of a gyroscope with sufficient dynamic range. A combat aircraft may have roll rates in excess of $300°/s$, and the acceptable drift rate was shown in the last section to be about $0\cdot01°/h$. The dynamic range requirement is therefore a staggering value of 10^8. Acceptable gyroscopes in several forms (ring laser, tuned rotor etc.) are now available, and give rise to interesting mathematical modelling problems in their own right. However, it is the computer equivalent of the mechanical gimbal system that will be discussed in this section.

Basically, it is the direction cosine matrix (DCM) of eqn. (5.1) which distinguishes the aircraft axis set $(Oxyz)$ from the tangent plane set $(O_0x_0y_0z_0)$. However, as formulated in (5.1) it is a function of the Euler angles Φ, Θ and Ψ. It has already been mentioned that these suffer from the mathematical equivalent or gimbal lock [remarks after eqn. (5.2)]. A further disadvantage is the computational complexity of generating trignometrical functions.

A popular alternative is to describe attitude by a four-parameter system, sometimes called Euler symmetrical parameters, Rodrigues parameters, Caley–Klein parameters or quaternion representation.[9] Let the four parameters be e_0, e_1, e_2 and e_3. Then the equivalent of eqn. (5.2) is

$$\begin{bmatrix} \dot{e}_0 \\ \dot{e}_1 \\ \dot{e}_2 \\ \dot{e}_3 \end{bmatrix} = \frac{1}{2} \begin{bmatrix} -e_1 & -e_2 & -e_3 \\ e_0 & -e_3 & e_2 \\ e_3 & e_0 & -e_1 \\ -e_2 & e_1 & e_0 \end{bmatrix} \begin{bmatrix} P \\ Q \\ R \end{bmatrix} \tag{5.88}$$

Since the three Euler angles are replaced by four parameters, there is obviously a redundancy equation. This turns out to be

$$e_0^2 + e_1^2 + e_2^2 + e_3^2 = 1 \qquad (5.89)$$

The DCM in terms of the parameters is

$$\begin{bmatrix} e_0^2 + e_1^2 - e_2^2 - e_3^2 & 2(e_1 e_2 - e_0 e_3) & 2(e_0 e_2 + e_1 e_3) \\ 2(e_0 e_3 + e_1 e_2) & e_0^2 - e_1^2 + e_2^2 - e_3^2 & 2(e_1 e_3 - e_0 e_1) \\ 2(e_1 e_3 - e_0 e_2) & 2(e_0 e_1 + e_2 e_3) & e_0^2 - e_1^2 - e_2^2 + e_3^2 \end{bmatrix} \qquad (5.90)$$

and can be used directly in eqn. (5.1). Apart from representing attitude without any singularities, the parameters are particularly suited to digital computation. Their value always lies in the interval $[-1 \ 1]$, so fixed point arithmetic can be used, and the constraint eqn. (5.89) helps (5.88) to be integrated in an accurate and stable way.

The main disadvantage is that they do not enable attitude to be visualized by the human observer, but the Euler angles may be deduced by comparing the terms in (5.90) with those in (5.1). Thus from element (31)

$$\sin \Theta = -2(e_1 e_3 - e_0 e_2)$$

Dividing element (21) by element (11) gives

$$\tan \Psi = 2(e_0 e_3 + e_1 e_2)/(e_0^2 + e_1^2 - e_2^2 - e_3^2)$$

Dividing element (32) by element (33) gives

$$\tan \Phi = 2(e_0 e_1 + e_2 e_3)/(e_0^2 - e_1^2 - e_2^2 + e_3^2)$$

The equations describing the behaviour of a strapdown INU explicitly involve the description of the aircraft motion because the sensors are aligned along the aircraft axes. Obviously both models should be in the same mathematical language, which is the additional justification mentioned at the beginning of this section, for writing the IN equations in a matrix form rather than vector-mechanics form.

The earth reference axes stored in the computer should also align with the aircraft axes at some simple attitude. This is why $O_0 z_0$ down has been chosen. The choice of $O_0 x_0$ pointing east was made because the aircraft moves east at $\Omega R \cos \lambda$ when it is geographically stationary.

However, there is some argument for $O_0 x_0$ to point north because the azimuth angle then lines up with compass card conventions.

5.7.5 Mixed navigation systems

As a final example of mathematical modelling in aerospace systems, the problem of mixing information generated in the INU with data from other sources will be considered. This usually arises because radio navigation aids such as Omega (a very low frequency phase comparison system) or doppler velocity measuring systems provide essentially the same information, but corrupted with different errors from those characteristic of the INU. By mixing the various sources of information in a Kalman filter a considerably better performance can be obtained than from any one navigator. The INU itself often contains a Kalman filter anyway to identify and compensate for the various error terms, and to align the platform after switch-on.

Most Kalman filter implementations are discrete-time and based on error states [such as eqn. (5.87)]. However for the sake of a simple illustration, the problem of mixing the height obtained from the INU, h_i and the height signal from the air data computer (barometric) will be considered.

From eqn. (5.77), $\dot{h} = -V_z$, and from eqn. (5.84) and (5.85)

$$\dot{V}_z = -2V_x\Omega \cos \lambda - \frac{1}{R+h}(V_x^2 + V_y^2) + a_z'$$ (5.91)

Assume that errors in the INU values of V_x, V_y and h plus accelerometer and gravitational compensation errors combine to form a random disturbance α.

Thus (5.91) can be written

$$\ddot{h} = -\dot{V}_{zi} + \alpha$$ (5.92)

where $\dot{V}_{zi}$ is the value of the right-hand side of (5.91) as calculated in the INU. Let $h = x_1$ and $\dot{h} = x_2$. Thus

$$\begin{bmatrix} \dot{x}_1 \\ \dot{x}_2 \end{bmatrix} = \begin{bmatrix} 0 & 1 \\ 0 & 0 \end{bmatrix} \begin{bmatrix} x_1 \\ x_2 \end{bmatrix} + \begin{bmatrix} 0 \\ -1 \end{bmatrix} \dot{V}_{zi} + \begin{bmatrix} 0 \\ 1 \end{bmatrix} \alpha$$ (5.93)

The air data computer produces a height estimate

$$h_a = h + \beta$$ (5.94)

where β represents the error in barometric data. Let $h_a = y$, and (5.94) becomes

$$y = \begin{bmatrix} 1 & 0 \end{bmatrix} \begin{bmatrix} x_1 \\ x_2 \end{bmatrix} + \beta$$ (5.95)

Equations (5.93) and (5.95) can be used as the state and output equations of a Kalman filter problem. For simplicity assume that α and β are white noise disturbances (unlikely to be valid assumptions in practice), and let

$$\epsilon\alpha^2 = q\delta \quad \text{and} \quad \epsilon\beta^2 = r\delta$$

The standard theory[4] shows that the steady-state Kalman filter is

$$\dot{\hat{x}} = A\hat{x} + Bu + P(y - C\hat{x} - D\hat{u})$$ (5.96)

where $\wedge$ means 'estimate of'.

The *filter gain matrix* P is given by $P = GC^T R^{-1}$, and G, the *error covariance matrix* defined by

$$G = \epsilon(x - \hat{x})(x - \hat{x})^T$$

can be found from the solution of the quadratic matrix equation (sometimes called the algebraic Riccati equation)

$$AG + GA^T - GC^T R^{-1} CG + Q = 0$$

In this case

$$R = r, \quad Q = \begin{bmatrix} 0 & 0 \\ 0 & q \end{bmatrix} \quad A = \begin{bmatrix} 0 & 1 \\ 0 & 0 \end{bmatrix} \quad \text{and} \quad C = \begin{bmatrix} 1 & 0 \end{bmatrix}$$

Hence

$$G = \begin{bmatrix} \sqrt{2}q^{1/4}r^{3/4} & \sqrt{qr} \\ \sqrt{qr} & \sqrt{2}q^{3/4}r^{1/4} \end{bmatrix}$$

and

$$P = \begin{bmatrix} p_1 \\ p_2 \end{bmatrix} = \begin{bmatrix} \sqrt{2}(q/r)^{1/4} \\ (q/r)^{1/2} \end{bmatrix} \tag{5.97}$$

Equation (5.96) written in full is

$$\begin{bmatrix} \dot{\hat{h}} \\ \dot{\hat{\dot{h}}} \end{bmatrix} = \begin{bmatrix} \hat{\dot{h}} \\ 0 \end{bmatrix} + \begin{bmatrix} 0 \\ -\dot{V}_{zi} \end{bmatrix} + \begin{bmatrix} p_1(h_a - \hat{h}) \\ p_2(h_a - \hat{h}) \end{bmatrix} \tag{5.98}$$

The block diagram is shown in Fig. (5.15).
The first element of the transfer function matrix of equation (5.98) is

$$\bar{h} = \frac{(-\dot{V}_{zi}) + (p_1 s + p_2)\bar{h}_a}{s^2 + p_1 s + p_2} \tag{5.99}$$

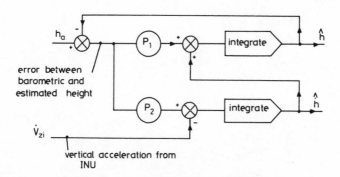

Fig. 5.15 *Block diagram of Kalman filter for h*

showing that the Kalman filter is acting as a so-called complementary filter with zero position and velocity-lag errors. As a matter of interest, substituting for p_1 and p_2 from eqn. (5.97) shows that the damping ratio of (5.99) is $1/\sqrt{2}$ for all disturbance strengths q and r.

5.8 References

1 McRUER, D., ASHKENAS, I., and GRAHAM, D.: 'Aircraft Dynamics and Automatic Control' (Princeton University Press, 1973)
2 BABISTER, A. W.: 'Aircraft Stability and Control' (Pergamon Press, 1961)
3 BEKEY, G. A., and KARPLUS, W. J.: 'Hybrid Computation' (John Wiley, 1968)
4 KWAKERNAAK, H., and SIVAN, R.: 'Linear Optimal Control Systems' (Wiley-Interscience, 1972)
5 FLYING QUALITIES OF PILOTED AIRPLANES: U.S. Government Printing Office MIL-F-8785B(ASG), 1969
6 TOBIE, H. N., ELLIOTT, E. M., and MALCOM, L. G.: 'A New Longitudinal Handling Quality Criterion', Proc. Nat. Aerosp. Elect. Conf., May 1966, Dayton, Ohio
7 McRUER, D. T. and KRENDEL, E. S.: 'Mathematical Models of Human Pilot Behaviour NATO Advisory Group for Aerospace Research and Development', AGARD-AG-188, January, 1974
8 ADAMS and HATCH: *Journal of Aircraft AIAA*, 1971, **8**, pp. 319–324
9 VANBRONKHORST, A.: 'Strapdown System Algorithms', NATO Advisory Group for Aerospace Research and Development AGARD-LS-95 pp. 3-1 to 3-22, May, 1978

Chapter 6

Marine systems

D. A. Linkens

6.1 Introduction

Marine vehicles are required to accomplish certain operating objectives in the environment of the sea. Within each vehicle there may be a number of subsystems, such as the power plant (propulsion and energy conversion), the control system (for ship steering), the cargo handling system and numerous military missions involving search-and-destroy systems. Each subsystem requires dynamic modelling and design to operate within its subenvironment 'optimally', together with satisfactory overall performance and survival of the integrated vehicle. In this chapter the modelling and control of one of the marine subsystems, being that of ship steering and manoeuvring, is considered.

Within the class of marine vehicles a number of different types can be distinguished. These include torpedoes, submarines and submersibles, military surface ships, commercial and fishing vessels, hydrofoil boats, hovercraft, towed oceanographic instruments and barges. Clearly, each of these categories have their own particular requirements and problems in dynamic modelling. For underwater vehicles, motion in three dimensions is involved giving similar considerations to those for the study of aircraft. For surface ships the motion can usually be simplified to that of one plane for horizontal manoeuvring. Thus, yaw response can be considered without coupling from heeling and pitching motion, unlike aircraft dynamics. This simplification of dynamics is accompanied by problems, however, particularly for large commercial ships. Thus, for relatively slow-moving ships performing large-angle manoeuvres the dynamics become nonlinear, and linearized mathematical models and autopilot controller design become invalid. In this chapter we shall concentrate on surface ship manoeuvring, emphasising the nonlinear aspects.

Having concentrated on the surface ship steering subsystems a number of considerations still need to be made. The operating requirements and specification must be established, such as the degree of manoeuvrability, the accuracy of course-keeping, together with the associated fuel costs and survivability of the vessel. These requirements must be considered under environmental conditions involving such things as surface gravity waves, water currents and turbulence, wind perturbations and the effects of shallow water and restricted channels. The dynamic modelling will involve a determination of the model structure, both in linearized and nonlinear form. This leads to an evaluation of the so-called hydrodynamic coefficients under different sea and wind conditions. The linearized equations of motion can then be analysed to determine the level of stability of the vessel. If this is unsatisfactory, improvements can be sought by changing inherent characteristics of the ship via such things as addition of stabilizing surfaces, increase in rudder size, additions of fixed appendages, e.g. bilge keels, or a major design change. Alternatively, or in addition to these changes, autopilot controllers can be designed and implemented. This pattern will be approximately followed in succeeding sections.

6.2 Analytical modelling of surface ship dynamics

A ship at sea can move in all the six degrees of freedom of motion—translation along three orthogonal axes and rotation about each of the three axes. The choice of an axes system which simplifies subsequent development is such that the origin should be at the point of symmetry. Most marine vehicles have a plane of symmetry defining two axes, with the third axis being perpendicular to that plane. The axes chosen are shown in Fig. 6.1. Abkowitz[1] points out that using axes of symmetry simplifies the hydrodynamic forces and the equations of motion through the fact that such axes are usually parallel to principal axes of inertia. Other advantages in this choice involve the model testing methods[2] and the problems caused by shifting centres of gravity.[3]

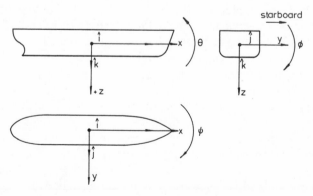

Fig. 6.1 *Body axes notation for ship steering*

Applying Newton's Laws of Motion to a body moving in a fluid gives the following very general relationship

Forces/moments = function   {properties of the body,
properties of motion
(orientation and dynamics),
properties of the fluid}.

The forces are determined from

$$\mathbf{F} = \frac{d}{dt}\,(\text{momentum}) = \frac{d}{dt}\,(m\mathbf{U}_G)$$

$$= m\frac{d}{dt}\,(\mathbf{U}_G) \qquad \text{for constant mass}$$

where subscript G refers to an origin at the centre of gravity. For an origin not at the centre of gravity of the body and in a system of axes fixed in and moving with the vehicle

$$\mathbf{U}_G = \mathbf{U}_a + \mathbf{\Omega} \times \mathbf{R}_G$$

where $\mathbf{U}_a$ is the velocity of the origin in space, $\mathbf{\Omega}$ is the angular velocity about the origin and $\mathbf{R}_G$ is the vector distance of the centre of gravity from the origin. However, since the origin is on the surface of the earth, and the earth rotates, then

$$\mathbf{U}_a = \mathbf{U} + \mathbf{\Omega}_e \times \mathbf{R}_p$$

where $\mathbf{U}$ is the geographical velocity of the body, $\mathbf{\Omega}_e$ is the angular velocity of the earth, and $\mathbf{R}_b$ is the radius vector from the earth's centre to the vehicle. The force equation thus becomes

$$\mathbf{F} = m\frac{d}{dt}\,(\mathbf{U} + \mathbf{\Omega}_e \times \mathbf{R}_b + \mathbf{\Omega} \times \mathbf{R}_G)$$

Expansion of the $\mathbf{\Omega}_e \times \mathbf{R}_b$ term produces the Coriolis force and centripetal acceleration due to rotation of the earth. For surface ships these terms are negligible, leaving the force equation

$$\mathbf{F} = m\frac{d}{dt}\,(\mathbf{U} + \mathbf{\Omega} \times \mathbf{R}_G)$$

Substituting the following defined quantities in the above expression

$$\mathbf{U} = iu + jv + kw$$
$$\mathbf{\Omega} = ip + jq + kr$$
$$\mathbf{R}_G = ix_G + jy_G + kz_G$$
$$\mathbf{F} = iX + jY + kZ$$

and noting the rotation relationships in pitch θ (about y axis), yaw ψ (about z axis) and roll ϕ (about x axis)—see Ref. 1, App. 1, pp. 5, 6.

$$\frac{di}{dt} = i0 + jr - kq$$

$$\frac{dj}{dt} = -ir + j0 + kp$$

$$\frac{dk}{dt} = iq - jp + k0$$

the following force equations result

$$X = m[\dot{u} + qw - rv - x_G(q^2 + r^2) + y_G(pq - \dot{r}) + z_G(pr + \dot{q})] \qquad (6.1)$$

$$Y = m[\dot{v} + ru - pw - y_G(r^2 + p^2) + z_G(qr - \dot{p}) + x_G(qp + \dot{r})] \qquad (6.2)$$

$$Z = m[\dot{w} + pv - qu - z_G(p^2 + q^2) + x_G(rp - \dot{q}) + y_G(rq + \dot{p})] \qquad (6.3)$$

Considering the forces arising from the earth's rotation to be negligible for marine vehicle dynamics the equivalent moment equation is

$$\mathbf{M}_G = \frac{d}{dt} \text{(angular momentum)}$$

For an origin not at the centre of gravity the moment expression becomes

$$\mathbf{M} = \mathbf{M}_G + \mathbf{R}_G \times \mathbf{F} = \mathbf{M}_G + \mathbf{R}_G \times m \frac{d}{dt}(\mathbf{U}_G)$$

For axes parallel to the principal axes of inertia the angular momentum about G is given by

$$\text{(angular momentum)}_G = iI_x p + jI_y q + kI_z r - m\mathbf{R}_G \times (\mathbf{\Omega} \times \mathbf{R}_G)$$

Substitution of vector components and reduction of terms leads to the following moment equations (see Ref. 1, App. 1, pp. 11, 13)

$$K = I_x\dot{p} + (I_z - I_y)qr + m[y_G(\dot{w} + pv - qu) - z_G(\dot{u} + ru + pw)] \qquad (6.4)$$

$$M = I_y\dot{q} + (I_x - I_z)rp + m[z_G(\dot{u} + qw - rv) - x_G(\dot{w} + pv - qu)] \qquad (6.5)$$

$$N = I_z\dot{r} + (I_y - I_x)pq + m[x_G(\dot{v} + ru - pw) - y_G(\dot{u} + qw - rv)] \qquad (6.6)$$

In these equations the terms $(qw - rv)$, $(ru - pw)$ and $(pv - qu)$ represent centripetal acceleration, and the terms $(I_z - I_y)qr$, $(I_x - I_z)rp$ and $(I_y - I_x)pq$ represent gyroscopic effects. The term involving x_G, y_G and z_G represent acceleration of the centre of gravity relative to the origin.

The forces and moments given in eqns. (6.1) to (6.6) are balanced by hydrostatic and hydrodynamic quantities which are functions of many variables.

Thus one can write

F, M

$$= f\{x_0, y_0, z_0, \phi, \theta, \psi, u, v, w, p, q, r, \dot{u}, \dot{v}, \dot{w}, \dot{p}, \dot{q}, \dot{r}, n, \dot{n}, \delta, \dot{\delta}, \ddot{\delta} \ldots\}$$

In the above expression n refers to the rotation for all rotating effectors such as main propulsion propellor, thrusters, etc., and δ refers to the deflection of all control surfaces such as rudders, elevators, propellor blade pitch etc. The variables x_0, y_0, z_0, ϕ, θ, and ψ deal with orientation in space and represent hydrostatic effects. These hydrostatic effects are calculable by theory, whereas the remaining hydrodynamic forces and moments are not. It is necessary to expand the hydrodynamic components via a Taylor expansion about a suitable initial condition which is usually taken to be straight-ahead motion at constant speed giving $u = u_0$, $v_0 = w_0 = p_0 = q_0 = r_0 = \delta_0 = \dot{\delta}_0 = 0$. Taylor expansion is limited to analytic functions, and hydrodynamic forces are assumed to be of this nature for the linear and third-order terms. Second-order terms can be accounted for using the form $u|u|$ and $v|v|$ etc.

In the linearized equations of motion only the linear terms in the Taylor expansion are retained. For a dynamically stable ship, linear theory holds true for moderate manoeuvres, and non-linear terms become necessary only for tight manoeuvres. For a dynamically unstable ship, however, higher-order terms are a requisite for determining manoeuvring properties. Linear terms in the Taylor expansion contain a partial derivative given for example by

$$\left(\frac{\partial N}{\partial v}\right)_0$$

where the derivative is taken at the original condition of $u = u_0$, $v = \dot{v} = r = \dot{r} = 0$. These so-called hydrodynamic derivatives are normally written in the following notation

$$\left(\frac{\partial N}{\partial v}\right)_0 \equiv N_v$$

Equations (6.1) to (6.6) together with the hydrostatic and hydrodynamic terms referred to above represent motion with six degrees of freedom. In horizontal plane manoeuvring of surface ships it is assumed that there is no cross-coupling with heave, roll and pitch of the vehicle. Linearizing the force and moment eqns. (6.1) to (6.6) and retaining only the relevant hydrodynamic derivatives leads to the following equations of motion for horizontal manoeuvring in deep unrestricted water at constant speed

$$(m - Y_{\dot{v}})\dot{v} = Y_v v + (Y_{\dot{r}} - m x_G)\dot{r} + (Y_r - mu)r + Y_\delta \delta \tag{6.7}$$

$$(I_z - N_{\dot{r}})\dot{r} = N_v v + (N_{\dot{v}} - m x_G)\dot{v} + (N_r - m x_G u)r + N_\delta \delta \tag{6.8}$$

Equations (6.7) and (6.8) can be rewritten in matrix form as

$$
\begin{pmatrix} m - Y_{\dot{v}} & mx_G - V_{\dot{r}} & 0 \\ mx_G - N_{\dot{v}} & I_z - N_{\dot{r}} & 0 \\ 0 & 0 & 1 \end{pmatrix} \begin{pmatrix} \dot{v} \\ \dot{r} \\ \dot{\psi} \end{pmatrix}
$$

$$
= \begin{pmatrix} Y_v & Y_r - mu & Y_{\psi}^w \\ N_v & N_r - mx_G u & N_{\psi}^w \\ 0 & 1 & 0 \end{pmatrix} \begin{pmatrix} v \\ r \\ \psi \end{pmatrix} + \begin{pmatrix} Y_\delta \\ N_\delta \\ 0 \end{pmatrix} \delta + \begin{pmatrix} F_1 \\ F_2 \\ 0 \end{pmatrix} \quad (6.9)
$$

where the force and moment due to wind disturbances are represented by $Y^w = Y_0^w + Y_{\psi}^w \psi$ and $N^w = N_0^w + N_{\psi}^w \psi$, and the initial wind load is given by $F_1 = Y_0^w - Y_\delta \delta_0$ and $F_2 = N_0^w - N_\delta \delta_0$.

Rearranging eqn. (6.9) gives the following transfer function relating yaw to rudder deflection

$$
G_{\psi\delta} = \frac{K(1 + T_3 s)}{s(T_1 T_2 s^2 + (T_1 + T_2)s + 1)} \quad (6.10)
$$

Equation (6.10) can be rewritten[4] as

$$
\dddot{\psi} + \left(\frac{1}{T_1} + \frac{1}{T_2} \right) \ddot{\psi} + \frac{1}{T_1 T_2} \dot{\psi} = \frac{K}{T_1 T_2} (T_3 \dot{\delta} + \delta) \quad (6.11)
$$

where the time constants and gain are related to the hydrodynamic derivatives by the expressions

$$
\frac{K}{T_1 T_2} = \frac{N_v Y_\delta - Y_v N_\delta}{(Y_{\dot{v}} - m)(N_{\dot{r}} - I_z) - (Y_{\dot{r}} - mx_G)(N_{\dot{v}} - mx_G)}
$$

$$
T_3 = \frac{(N_{\dot{v}} - mx_G)Y_\delta - (Y_{\dot{v}} - m)N_\delta}{N_v Y_\delta - Y_v N_\delta}
$$

$$
\frac{1}{T_1 T_2} = \frac{Y_v(N_r - mx_G u_0) - N_v(Y_r - mu_0)}{(Y_{\dot{v}} - m)(N_{\dot{r}} - I_z) - (Y_{\dot{r}} - mx_G)(N_{\dot{v}} - mx_G)}
$$

$$
\left(\frac{1}{T_1} + \frac{1}{T_2} \right) = \frac{\begin{aligned} (Y_{\dot{v}} - m)(N_r - mx_G u_0) + (N_{\dot{r}} - I_z)Y_v \\ - (Y_{\dot{r}} - mx_G)N_v - (N_{\dot{v}} - mx_G)(Y_r - mv_0) \end{aligned}}{(Y_{\dot{v}} - m)(N_{\dot{r}} - I_z) - (Y_{\dot{r}} - mx_G)(N_{\dot{v}} - mx_G)}
$$

The transfer function of eqn. (6.10) has been approximated by Nomoto[5] to

$$
G_{\psi\delta} \approx \frac{K}{s(1 + T_s)} \quad (6.12)
$$

where the effective time constant is $T = T_1 + T_2 - T_3$. This simplified dynamic has been successfully used for analysing steering problems for stable ships, but is inadequate for unstable ships.

At this point it is convenient to explain what is meant by stable and unstable ships in this context. Introducing the Laplace operators into the equations of motion (6.7) and (6.8) enables them to be written as

$$A \begin{pmatrix} \Delta u \\ v \\ r \end{pmatrix} = \begin{pmatrix} 0 \\ 0 \\ 0 \end{pmatrix} \quad \text{where} \quad A = \begin{pmatrix} a_{11} & a_{12} & a_{13} \\ a_{21} & a_{22} & a_{23} \\ a_{31} & a_{32} & a_{33} \end{pmatrix}$$

and $a_{11} = (X_{\dot{u}} - m)s + X_u$ etc.

The stability of the linearized vehicle can be determined by the sign of the roots of $|A| = 0$. By virtue of the choice of symmetrical body axes, it can be shown (e.g. Ref. 1) that the resulting relationships are simplified since X_v, $X_{\dot{v}}$, X_r and $X_{\dot{r}}$ are zero. As a result, elements a_{12} and a_{13} are zero, and the roots are given by

$$a_{11}(a_{22}a_{33} - a_{23}a_{32}) = a_{11}(As^2 + Bs + C) = 0 \qquad (6.13)$$

where

$$A = (Y_{\dot{v}} - m)(N_{\dot{r}} - I_z) - (Y_{\dot{r}} - mx_G)(N_{\dot{v}} - mx_G)$$

$$B = (Y_{\dot{v}} - m)(N_r - mx_G u_0) + (N_{\dot{r}} - I_z)Y_v$$

$$\quad - (Y_{\dot{r}} - mx_G)N_v - (N_{\dot{v}} - mx_G)(Y_r - mu_0)$$

$$C = Y_v(N_r - mx_G u_0) - N_v(Y_r - mu_0)$$

The roots of (6.13) are thus given by

$$\alpha_1 = -\frac{X_u}{(X_{\dot{u}} - m)}$$

$$\alpha_{2,3} = \frac{-B \pm \sqrt{B^2 - 4AC}}{2A}$$

For marine vehicles both X_u and $X_{\dot{u}}$ are negative, and $X_{\dot{u}}$ is about 5 to 10% of the mass for surface ships, and hence the root α_1 is always negative indicating stability. The real parts of the roots α_2 and α_3 will be negative, and therefore indicate a stable ship provided that both $AB > 0$ and $AC > 0$. It can be shown (e.g. Refs. 1 and 3) that, for ships, both A and B are always positive and hence the stability criterion reduces to $C > 0$. Thus the condition for dynamic stability in straight-line motion is

$$Y_v(N_r - mx_G u_0) - N_v(Y_r - mu_0) > 0 \qquad (6.14)$$

The terms in (6.14) are major parameters in naval architecture design. Both Y_v and N_r are negative in value with bow and stern effects additive in nature. Thus the term $Y_v(N_r - mx_G u_0)$ is a large-valued positive quantity. For Y_r, bow and stern effects fight each other, so that it can be positive or negative. However, $Y_r < mu_0$ and hence $(Y_r - mu_0)$ is always negative. Thus, dynamic stability is assured if N_v is positive, which occurs when the stern effects predominate over bow effects. While the majority of fast, sleek ships are dynamically stable, this is not the case for large tankers. The manner in which instability is exhibited in ship dynamics testing will be described in the next section.

The problem of non-linear dynamics becomes important when dealing with large commercial vessels undergoing heavy manoeuvres. The determination of a valid non-linear mathematical model has been approached from two aspects. The first approach incorporates the non-linear force terms and non-linear components in the Taylor expansion of the hydrodynamic terms. Clearly, the measurement of such non-linear derivatives will be important in this approach and will be considered in the section on testing. The conventional X force equation in linearized form is

$$X_{\dot{u}}u + X_u\,\Delta u + X_{\dot{v}}\dot{v} + X_v v + X_{\dot{r}}\dot{r} + X_r r + X_\delta \delta = m\dot{u}$$

As mentioned earlier, the derivatives X_v, X_r, X_δ, $X_{\dot{v}}$ and $X_{\dot{r}}$ are zero due to body symmetry, so that the forces can be represented by even functions when allowing for non-linear effects e.g.

$$X(v) = a_2 v^2 + a_4 v^4 + a_6 v^6$$

Since v, r and δ are all varying simultaneously, the force should be represented by an even power expansion of the effective sum of the parameters, i.e.

$$X(r, v, \delta) = (k_1 r + k_2 v + k_3 \delta)^2 + (k_1 r + k_2 v + k_3 \delta)^4 \cdots$$

In ship modelling it is considered sufficient to retain terms up to third order, leaving in this case non-linear terms of

$$r^2,\ v^2,\ \delta^2,\ rv,\ r\delta,\ v\delta$$

As a result, the expected form of the non-linear equation for x is

$$a\dot{u} + b\,\Delta u + c(\Delta u)^2 + d(\Delta u)^3 + ev^2 + fr^2 + g\delta^2 + hrv + jr\delta + e_1 v^2\,\Delta u$$
$$+ f_1 r^2\,\Delta u + g_1 \delta^2\,\Delta u + h_1 rv\,\Delta u + j_1 r\delta\,\Delta u + k_1 v\delta\,\Delta u = 0$$

For the Y and N equations the terms involving r, v and δ will be odd functions, requiring to be represented by

$$Y(v) = b_1 v + b_3 v^3 + b_5 v^5 + \cdots$$

In this case power series expansion allowing for simultaneous changes in r, v and δ and including terms up to the third order involves components in r, v, δ, r^3, v^3, δ^3, $r^2\delta$, $\delta^2 r$, $r^2 v$, $v^2 r$, $\delta^2 v$, $v^2\delta$, $rv\delta$. The resulting non-linear equation for Y takes the form

$$A\dot{v} + B\dot{r} + Cv + Dr + E\delta + Fv^3 + Gr^3 + H\delta^3 + Ir^2\delta + J_1\delta^2 r + Kv^2 r$$
$$+ Mr^2 v + N_1\delta^2 v + Pv^2\delta + P_1 vr\delta + Qv\,\Delta u + Rv(\Delta u)^2 + Sr\,\Delta u$$
$$+ Tr(\Delta u)^2 + V\delta\,\Delta u + W\delta(\Delta u)^2 + Y_0[1 + A_1\,\Delta u + B_1(\Delta u)^2] = 0$$

A similar formidable relationship results for the moment equation N. While these coefficients can be determined from scale-model experiments using sophisticated equipment it obviously requires a vast amount of measurement with doubtful usefulness in view of problems over scaling.

The second method of approach to non-linear modelling is more of a grey-box nature. Returning to the single third-order differential equation of (6.11) for manoeuvring it has been suggested[6, 4] that the coefficients $[(1/T_1) + (1/T_2)]$, $K/T_1 T_2$ and T_3 all remain fairly constant during manoeuvres for a given ship at a constant speed. In contrast, $1/T_1 T_2$ is far from being constant. Bech[4] thus suggested that the non-linear behaviour should be completely assigned to this term, giving the following equation

$$\dddot{\psi} + \left(\frac{1}{T_1} + \frac{1}{T_2}\right)\ddot{\psi} + \frac{K}{T_1 T_2} H(\psi) = \frac{K}{T_1 T_2}(T_3\dot{\delta} + \delta) \qquad (6.15)$$

In an attempt to amalgamate these two approaches Clarke[7] made use of the fact that side-slipping behaviour closely follows that in yaw, and in steady turning there is an almost linear relationship between side-slip velocity and yaw-rate. Using this relationship, side-slip velocity can be eliminated from the non-linear terms of the full equations of motion, thus producing a single differential equation equivalent to (6.15) given by

$$A_0\ddot{r} + (A_1 + A_2 r + A_3 r^2)\dot{r} + (A_4 + A_5 r + A_6 r^2)r$$
$$+ (A_7 + A_8\delta^2)\dot{\delta} + (A_9 + A_{10}\delta^2)\delta + A_{11} = 0 \qquad (6.16)$$

where the coefficients A_0 to A_{11} are functions of the hydrodynamic coefficients. The Nomoto and Clarke models can be related by noting that

$$A_0/A_4 = T_1 T_2$$
$$A_1/A_4 = T_1 + T_2$$
$$A_7/A_4 = -kT_3$$
$$A_9/A_4 = -k$$

Also, for steady turning, eqn. (6.16) reduces to

$$\delta = \frac{(A_4 + A_5 r + A_6 r^2)r + A_{11}}{A_9}$$

which can be considered as an analytic function equivalent to the $H(\psi)$ non-linear term of the Bech model [eqn. (6.15)].

An alternative approach to modelling beyond the conventional linear region employs functional analysis.[8] In this approach the slow-motion hydrodynamic derivatives based on quasi-steady fluid flow are replaced with a Volterra series specification which allows time-history effects of the flow to be considered. It was demonstrated that, in certain cases, ship-testing results which would normally have been assumed to require non-linear representations could be capable of accurate linear specification in this way.[9] Further analysis of these 'memory' effects[10, 11] suggest that their significance is small for large slow tankers, and that non-linear effects predominate.

A further non-linear model proposed for low-speed manoeuvring of large vessels in shallow water introduces a series of functions, of at most three dimensions each, giving access to graphical representations in three-dimensional spaces.[12] The equations are

$$m'_u(\dot{u} - vr) = X_h(u, v, r) + X_c(u, v, \delta) + X_s(W_u, W_v) + X_w(\xi, \tau, \phi)$$

$$m'_v(\dot{v} + ur) = Y_h(u, v, r) + Y_c(u, v, \delta) + Y_s(W_u, W_v) + Y_w(\xi, \tau, \phi) \qquad (6.17)$$

$$I'\dot{r} = N_h(u, v, r) + N_c(u, v, \delta) + N_s(W_u, W_v) + N_w(\xi, \tau, \phi)$$

where m'_u and m'_v include 'added mass' terms due to acceleration, suffix h denotes hull forces, suffix c denotes control forces (e.g. propeller and non-central rudder), suffix s denotes superstructure forces including wind effects, and suffix w indicates drift forces caused by wave action. ξ, τ and ϕ are the wave height, period and ship bearing, respectively.

6.3 Identification of surface ship dynamics

Experimental determination of parameters in the ship-steering models of the previous section falls into two categories. These comprise scale-model testing and full-size sea trials. In the former case the models are usually of a 'captive' nature with sophisticated tank and measurement facilities. This enables many parameters to be estimated, but suffers from the disadvantage of scaling problems. Such scaling problems are accentuated for surface ship studies where there is a water-to-air interface. In sea trials the amount of experimentation possible is limited due to expense of ship operation and instrumentation. The number of parameters which can therefore be estimated is correspondingly smaller.

6.3.1 Scale-model testing
The hydrodynamic derivatives occurring in the equations of motion cannot be calculated analytically and hence model tests are carried out in towing tanks and rotating arm tanks. A small sideways velocity can be superimposed on the model by towing a model at an angle to its direction of motion. This is called the 'oblique tow test' and is illustrated in Fig. 6.2. The values of Y_v and N_v can be determined from such tests, and can be non-dimensionalized giving values which are constant with speed.[3] In all model testing it is necessary to simulate the correct pattern of flow around the hull. For surface ships the wave patterns are made correct by equating the Froude numbers of the full-size and model situations. The resulting error in Reynolds number has to be compensated using experimentally determined graphs of non-dimensionalized drag coefficients versus Reynolds number for the particular

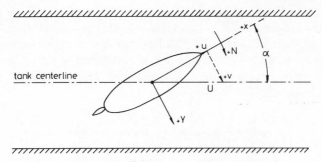

Fig. 6.2   *Schematic diagram for oblique tow model testing*

hull shape being studied. The propellers should be operating at the correct scaled thrust, and the rudder should be fixed in the undeflected position.

The rotating arm tank enables an angular velocity to be superimposed on an equilibrium forward speed, as shown in Fig. 6.3. In addition to the precautions required for a linear towing tank it is necessary to accelerate the model and collect the necessary data within one revolution of the rotating arm. This is necessary to avoid the model running in its own wake. Y_r and N_r can be determined from these tests although it should be noted that large ratios of R/L, where L is the length of the model, are required to enable the slope of the characteristic at small r to be determined to give the hydrodynamic derivative. If the model is ballasted so that its weight equals its buoyancy, and the centre of gravity of the model and the ship are equivalent, then force and moment measurements will provide the $(Y_r - mu_0)$ and $(N_r - mx_G u_0)$ terms in the stability criterion of eqn. (6.14). By making measurements over a wide range of v and r, some of the non-linear coefficients can be determined from these tests via curve-fitting.

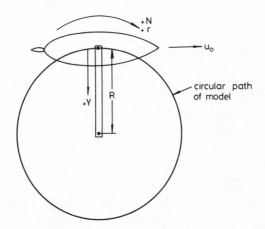

Fig. 6.3   *Schematic diagram for rotating arm tank model testing*

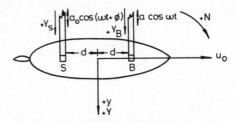

Fig. 6.4 *Principle of Planar Motion Mechanism constrained model testing*

Another form of constrained model testing uses a so-called Planar Motion Mechanism (PMM) pioneered by Gertler[13] and Goodman.[14] This can be used in a linear towing tank and enables the derivatives Y_r and N_r to be determined, and also the acceleration derivatives $N_{\dot v}$, $Y_{\dot r}$, $Y_{\dot v}$ and $N_{\dot r}$ to be found. The mechanism basically comprises two mechanical oscillating struts attached to the model which produce transverse oscillations on the model as it is towed along the tank at constant speed. The bow and stern oscillators have the same frequency and amplitude but with an adjustable phase angle between them, as shown in Fig. 6.4. Strain gauges are used to measure the forces at the bow and stern struts Y_B and Y_S. If the phase angle ϕ is set to 0 the model experiences pure transverse motion and Y_v and N_v can be determined from

$$Y_v = \frac{\partial Y}{\partial v} = \pm \frac{\text{out-of-phase amplitude of } (Y_B + Y_S)}{-a_0 \omega}$$

$$N_v = \frac{\partial N}{\partial v} = \pm \frac{\text{out-of-phase amplitude of } (Y_B - Y_S) \, d}{-a_0 \omega}$$

The frequency dependency of these derivatives can be found by varying ω, while the zero-frequency derivatives can be determined by the oblique tow test. The derivatives $Y_{\dot v}$ and $N_{\dot v}$ can be obtained from

$$Y_{\dot v} = \frac{\text{in-phase amplitude of } (Y_B + Y_S)}{-\omega^2 a_0}$$

$$N_{\dot v} = \frac{\text{in-phase amplitude of } (Y_B - Y_S) \, d}{-\omega^2 a_0}$$

Given proper ballasting, the in-phase components will give the stability terms $(Y_{\dot v} - m)$ and $(N_{\dot v} - mx_G)$ directly.

In order to obtain angular motion derivatives the model has to oscillate so that it is always tangential to its path. This can be achieved by setting the phase angle to

$$\phi/2 = \tan^{-1} \frac{\omega d}{u_0}$$

The derivatives are then determined as follows.

$$(Y_r - mu_0) = \pm \frac{\text{out-of-phase amplitude of } (Y_B + Y_S)}{-\psi_0 \omega}$$

$$(N_r - mx_G u_0) = \pm \frac{\text{out-of-phase amplitude of } (Y_B - Y_S)\, d}{-\psi_0 \omega}$$

$$(Y_{\dot{r}} - mx_G) = \pm \frac{\text{in-phase amplitude of } (Y_B + Y_S)}{-\psi_0 \omega^2}$$

$$(N_{\dot{r}} - I_z) = \pm \frac{\text{in-phase amplitude of } (Y_B - Y_S)\, d}{-\psi_0 \omega^2}$$

Some of the questions regarding the frequency-dependent nature of these PMM coefficients can be found in Refs. 15 and 9.

Large PMM testing facilities have been developed at the David Taylor Model Basin in America, principally used for submarine investigations,[14] and at the HyA Laboratory, Denmark for testing surface ship models of up to 6 m in length.[16]

6.3.2 Full-scale sea trials

Theoretical and model studies of ship manoeuvrability must always be supplemented by actual sea trials of the vessel. Over the years a number of standard manoeuvres and measurements have been established which enable some of the parameters in the mathematical models of Section 6.2 to be estimated. The classical acceptance trial manoeuvre is the 'hard-over turning circle' which is illustrated in Fig. 6.5. In naval arthitecture terms this is specified in terms of minimum 'reach', 'advance' and 'tactical diameter' for

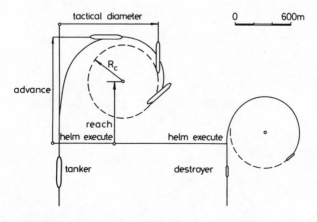

Fig. 6.5 *Typical turning circle tracks for a 300 m tanker and a 110 m destroyer*

maximum helm deflection. Typical values for tactical diameter are about 3·5 ship lengths for a tanker and 7 ship lengths for a destroyer at high speed. In terms of ship lengths the 'reach' of a tanker is about twice that of a destroyer indicating a larger effective time constant equivalent to the first order Nomoto model of eqn. (6.12). Although the actual turning circles for a tanker and destroyer are similar, the non-dimensional turning-path curvature L/R_c (or yaw rate $\dot{\psi}$) is much larger for a tanker, indicating a higher effective gain $K = \dot{\psi}/\delta$ [see eqn. (6.12)].

By repeating the above manoeuvre for different rudder angles and measuring the steady-state yaw rate in each case the spiral manoeuvre characteristic can be plotted. The rudder increments are usually 5° at large helms and 1° or 2° near midships. This characteristic clearly demonstrates whether the ship is stable or unstable. As shown in Fig. 6.6 a stable ship has only one yaw rate for a given rudder deflection. In contrast, an unstable ship has two values of yaw rate for a given small rudder deflection, and in particular this shows that such a ship will not steer a straight course for $\delta = 0$. It is, of course, possible to steer an unstable ship on a straight course either by a good helmsman or an autopilot. An example of this is shown in Fig. 6.7 for a simulator plus manual steering with the time course history of $\dot{\psi}$ superimposed on the steady-state spiral characteristic. The non-linear function $H(\dot{\psi})$ in the model of eqn. (6.15) by Bech can be determined from these spiral manoeuvres. In the case of a stable ship the normal Dieudonné spiral test can be used, while for an unstable ship the reversed spiral test of Bech[6] is applicable. In the reversed spiral test the characteristic is obtained by determining the rudder angle necessary to sustain a particular yaw rate.

Another type of manoeuvring trial is the Kempf zigzag test or z-manoeuvre. With the ship proceeding at speed with zero rudder, the rudder is deflected to,

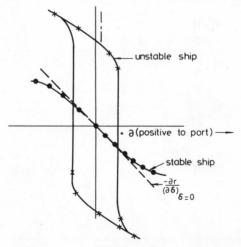

Fig. 6.6 *Typical spiral manoeuvre characteristics for a stable ship and an unstable ship*

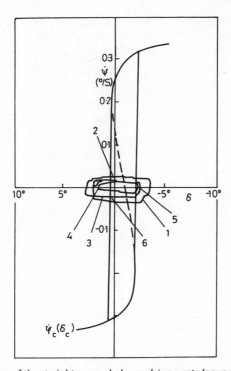

Fig. 6.7   *Time history of the straight course helm angle/ yaw rate locus of a 200,000 tdw tanker during manual control on a simulator, plotted on top of the steady state steering characteristics*
(From Aström, K. J., Källström, C. G., Norrbin, N. H., and Byström, L., 'The Identification of Linear Ship Steering Dynamics using Maximum Likelihood Parameter Estimation', SSPA Report 1920–1)

say, 20° to starboard and held in this position until the ship has changed its heading to 20° starboard. At this point the rudder is quickly deflected 20° to port and held there until the ship reaches a 20° port heading, at which point the cycle of events is repeated. The heading time-response due to this square-wave rudder input gives an indication of the ship dynamics. The overshoot in heading angle is related to the dynamic stability of the vessel. Bech and Wagner-Smitt[4] proposed a method of finding the parameters in the mathematical model of eqn. (6.15) based on a phase-plane construction from zigzag trial measurements. Figure 6.8 shows an example of typical zigzag responses together with the $H(\dot{\psi})$ characteristic and phase-plane portrait. Also shown on the figure are the points on the phase-plane used for determination of the parameters. If S is the slope of the phase trajectory, $\ddot{\psi}$ can be replaced by $S\dot{\psi}$ and eqn. (6.15) can be rewritten as

$$S\dot{\psi} = \frac{K}{T_1 T_2}(\delta - H(\dot{\psi})) - \left(\frac{1}{T_1} + \frac{1}{T_2}\right)\dot{\psi}$$

N.I. — N

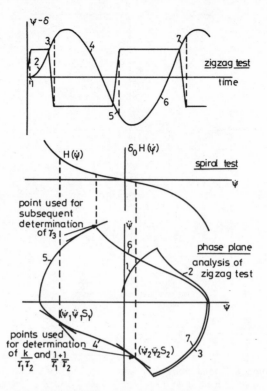

Fig. 6.8   *Parameter determination via phase-plane analysis of a zig-zag test*
(From Bech, M., and Smith, W. L., 'Analog Simulation of Ship Manoeuvres', HyA.
Report Hy-14)

For δ constant and known, and at any point on the trajectory $H(\dot\psi)$ being
known from a spiral test, the unknowns $(K/T_1 T_2)$ and $[(1/T_1) + 1/T_2)]$ can be
determined using data from two points $(\dot\psi_1, \ddot\psi_1, S_1)$ and $(\dot\psi_2, \ddot\psi_2, S_2)$ marked
in Fig. 6.8. The time constant T_2 can be obtained from the phase-plane trajec-
tory where the rudder is moved at constant rate $\dot\delta$. The end point of the
trajectory as marked is suitable since the rudder angle δ is defined there, and
a maximum value of $\ddot\psi$ is achieved making it easier to define the tangent at
that point. Because of the need to differentiate the yaw-rate signal to obtain
the phase-plane portrait this method is open to error caused by noisy measurement and requires careful filtering of signals to achieve sensible results. That
linear dynamics are unsuitable for representing the zigzag type of sea-trial can
be seen from Fig. 6.9 which shows a 20°/20° manoeuvre for a simulated
container ship where the non-linear contribution in the yaw acceleration $\ddot\psi$
response is very significant.

The use of system identification techniques to determine ship dynamics has
been increasing in recent years. A conventional least-squares method for estimating the two parameters in the model of eqn. (6.12) using the integrated

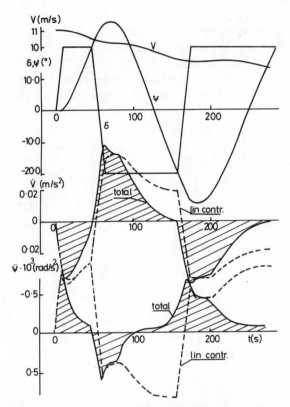

Fig. 6.9   *Total and linear contribution to sway and yaw accelerations in the mathematical modelling of a 20°/20° zig-zag manoeuvre with a twin-screw container ship*
(From Aström, K. J., Källström, C. G., Norrbin, N. H., and Byström, L., 'The Identification of Linear Ship Steering Dynamics using Maximum Likelihood Parameter Estimation', SSPA Report 1920–1)

response to a succession of ramp inputs was suggested by Nomoto.[17] Although sinusoidal and transient response methods have been reported for submarine investigations, these have not been common for full-size surface ship measurements. This is because precise inputs and large-course deviations are required together with long experimental trials.

An obvious alternative test input would be that of a PRBS perturbation. This has the advantage that the perturbations need only be small, thus making linearized dynamics more realistic and also simplifying the sea-trials procedures. PRBS techniques have been applied to ship-roll dynamics for a frigate with apparent success provided that the signal parameters were chosen carefully.[18]

Several parameter estimation schemes have been applied to determine dynamics from full-size ship trials. An iterative least-squares method was proposed[19] with the use of an extended Kalman filter when only certain of the state variables were measured. This method required continuous records to be available, meaning that short sampling periods (of about 1 second) would be necessary, giving large amounts of data. The method of maximum likelihood has been used by Källström and Aström[20-22] to fit a continuous-time ship model using discrete-time measurements. In this way long sampling periods (10 to 30 seconds) can be used, thus reducing the computational load. A similar procedure has been adopted for the ship *Esquilino*.[23] In this method the linearized matrix equations of motion are written as

$$
\begin{pmatrix} \dot{v} \\ \dot{r} \\ \dot{\psi} \end{pmatrix} = \begin{pmatrix} a_{11} & a_{12} & a_{13} \\ a_{21} & a_{22} & a_{23} \\ 0 & 1 & 0 \end{pmatrix} \begin{pmatrix} v \\ r \\ \psi \end{pmatrix} + \begin{pmatrix} b_{11} \\ b_{12} \\ 0 \end{pmatrix} \delta + \begin{pmatrix} f_1 \\ f_2 \\ 0 \end{pmatrix}
\tag{6.18}
$$

where f_1 and f_2 are disturbances and

$$
\begin{pmatrix} a_{11} & a_{12} & a_{13} & b_{11} & f_1 \\ a_{21} & a_{22} & a_{23} & b_{21} & f_2 \end{pmatrix} = \begin{pmatrix} m - Y_{\dot{v}} & mx_G - Y_{\dot{r}} \\ mx_G - N_{\dot{v}} & I_z - N_{\dot{r}} \end{pmatrix}^{-1}
$$

$$
\times \begin{pmatrix} Y_v & Y_r - mu_0 & Y_\psi^w & Y_\delta & F_1 \\ N_v & Nr - mu_0 x_G & N_\psi^w & N_\delta & F_2 \end{pmatrix}
$$

The transfer function relating heading angle to rudder is given by

$$
G(s) = \frac{b_1' s + b_2'}{s^3 + a_1' s^2 + a_2' s + a_3'}
\tag{6.19}
$$

where the parameters are interrelated by

$$a_1' = -a_{11} - a_{22}$$

$$a_2' = -a_{12}a_{21} + a_{11}a_{22} - a_{23}$$

$$a_3' = -a_{13}a_{21} + a_{11}a_{23}$$

$$b_1' = b_{21}$$

$$b_2' = a_{21}b_{11} - a_{11}b_{21}$$

This continuous-time model is represented by a discrete-time model of the form

$$
A(z^{-1})y_k = B(z^{-1})u_k + C(z^{-1})e_k
\tag{6.20}
$$

where z^{-1} is the unit delay operator, and A, B and C are polynomials given by

$$A(z^{-1}) = a_1 z^{-1} + a_2 z^{-2} + \cdots$$

$$B(z^{-1}) = b_1 z^{-1} + b_2 z^{-2} + \cdots$$

$$C(z^{-1}) = 1 + c_1 z^{-1} + c_2 z^{-2} + \cdots$$

It can be shown[22] that the five parameters of eqn. (6.19) can be identified from measurements of heading angle and rudder input. If sway velocity is also measured, three additional parameters can be identified, so that the eight parameters of eqn. (6.18) can be obtained. The full set of parameters involving the hydrodynamic derivatives cannot be determined by these input-output trials.

Three sets of experimental trials have been reported by Aström and Källström.[22] The first was on a container ship, the *Atlantic Song*, at 18 knots under fresh gale conditions. A sampling interval of 15 seconds was used for a 30-minute experiment. A PRBS signal was injected manually via a helmsman and comprised two periods of a 64-length sequence with a peak-to-peak variation of 10°. For a second-order model the following parameters were obtained using maximum likelihood estimation

$$a_1 = -1 \cdot 64 \pm 0 \cdot 05 \qquad c_1 = -0 \cdot 75 \pm 0 \cdot 10$$

$$a_2 = 0 \cdot 66 \pm 0 \cdot 05 \qquad c_2 = 0 \cdot 06 \pm 0 \cdot 10$$

$$b_1 = -0 \cdot 11 \pm 0 \cdot 03 \qquad \text{Loss} = 294 \cdot 5$$

$$b_2 = -0 \cdot 19 \pm 0 \cdot 04 \qquad \text{AIC} = 560 \cdot 5$$

where AIC is the Akaike Information Criterion[24] used to indicate the order of model necessary to accurately model the data. The AIC indicator showed little improvement in going from a second- to a third-order model. In fact, difficulty was experienced in making a third-order model converge. The results of identification using a second-order model are shown in Fig. 6.10 which indicate large measurement errors at times of 1200 and 1600 seconds. Using data from the first 1200 seconds, and with level and trend removal attempted, made little difference to the parameter estimates. Parameters a_1 and a_2 gave a zero for $A(z^{-1})$ close to -1, indicating an integration in the model transfer function as expected. The parameters obtained corresponded to a Nomoto first-order model [eqn. (6.12)] with a gain of $-0 \cdot 055$ s^{-1} and a time constant of 29·4 s, together with a pure time-delay of about 15 s (i.e. one sample lag). The time delay is introduced to model the lag in the rudder servo and produces a non-minimum phase characteristic, also noted in Ref. 18, but attributed there to hydrodynamic effects. Provided that the process noise was modelled [via $C(z^{-1})$] there was no significant improvement in going from a second- to a third-order continuous model. It was noted, however, that considerable differences in estimates of gain and time constants were obtained if the process noise was not modelled. A change from second- to third-order discrete-time model [eqn. (6.20)] gave whiter model residuals, but this was attributed to the effect of round-off errors in heading angle measurement.

A similar experiment was performed on a twin-screw car ferry *Bore 1* except that the analogue signals obtained were sampled at 1 Hz to give more accurate resolution of time constants which were of the order of 10 seconds.

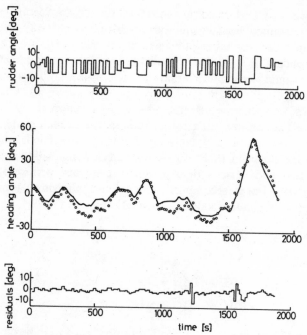

Fig. 6.10   *Identification of a second-order discrete model to 'Atlantic Song' PRBS data.
Heading measurements dots, heading model full line*
(From Aström, K. J., Källström, C. G., Norrbin, N. H., and Byström, L., 'The
Identification of Linear Ship Steering Dynamics using Maximum Likelihood
Parameter Estimation', SSPA Report 1920-1)

Using the PRBS results a third-order model of eqn. (6.19) type was fitted and
compared with a second-order Nomoto model fitted to zigzag manoeuvre
trials. It was concluded that the PRBS frequency excitation range and the
third-order model gave better modelling results for non-dimensional frequencies higher than $\omega = 0.5$.

Experiments were performed on two large oil tankers. A $10°/10°$ zigzag test
was analysed from the delivery trials for the *Fernström*. In this case it was
possible to calculate the sway velocity $\dot{v}$ from heading measurements and
Decca co-ordinates. Fitting of second- and third-order continuous transfer
functions again showed that the second-order approximation was a good
approximation only for low frequencies. As noted already, $\dot{v}$ measurements
allow further parameters to be estimated. Some of the hydrodynamic derivatives were estimated using these with initial values being estimated from
PMM measurements from a similar type of tanker. Using the $\dot{v}$ data the
estimates for a third-order continuous model of type eqn. (6.19) were very
different, giving K to be positive and a denominator time constant to be
negative. This indicated an unstable ship, showing that linear estimation becomes invalid for such manouvres.

PRBS tests were performed on another oil tanker, the *Sea Splendour*. In this case more instrumentation was included to give bow and stern sway velocities with a doppler log. Estimates of hydrodynamic derivatives from the trial data differed considerably from PMM model tests on a similar tanker. Correlation analysis indicated that the model was questionable, and this was attributed to non-linear effects during a course change in the middle of the experiment. While process noise modelling appeared to be unnecessary for the hydrodynamic derivatives estimation, this was not the case for gain and time constant estimation.

Comparisons were made between PMM model test results and zigzag trial parameter estimates from a Mariner-type cargo ship *Compass Island*. It was concluded that considerable errors occurred in hydrodynamic derivative estimation using zigzag trials because of the considerable non-linear effects during these manoeuvres. From these extensive trials it is clear that for commercial ship steering modelling, the assumption of a linear model is often invalid. While a linear model may prove satisfactory for autopilot design and control for course-keeping against wind and water disturbances, this is clearly not valid for manoeuvres involving course changes. Identification studies similar to those of Källström and Aström have been performed on an Italian oceanographic vessel, *Bannock*.[25]

6.4 Automatic steering control of ships

The modelling structures and identifications of the previous sections are important in the design of autopilots for ships. In this section a brief outline will be given of automatic control development in this area. The classical method of autopilot control is to use a PID controller (proportional, integral, derivative) and to tune the individual terms during sea trials. The disadvantage with using a fixed-term PID controller is that the ship parameters change significantly in different operating conditions, for example when manoeuvring in deep or shallow waters.

Aström[26] has shown, via sensitivity analysis on a simple ship model of a tanker, that a fixed PID controller can provide regulation of a ship which can be either inherently stable or unstable depending on operating conditions. It can provide reasonable damping over a wide operating range. Equivalent results were obtained for a container ship, *Manchester Challenge*.[27] A fixed gain regulator will, however, have a poor performance in many of the operating conditions. The performance criterion can be improved significantly by tuning the controller parameters.

On-line computer control of the *Bannock* was implemented using a minimum variance optimal control law based on a cost function of

$$ J = \sum_{n=0}^{N} y^2(n) + a \sum_{n=0}^{N} u^2(n) $$

where y is the heading error and u is the rudder action. About 25 hours of computer control trials were performed, seven of which could be partly compared with conventional PID control. Improved course-keeping was claimed for the optimal controller with satisfactory convergence to minimal heading error variance under widely differing sea conditions. The performance index can be chosen to reflect energy consumption, and Norrbin[28] showed that for small perturbations about a straight course the average increase in resistance due to steering could be approximated by

$$\frac{\Delta R}{R} = \mu(\psi^{-2} + \rho\delta^{-2})$$

where R is drag and ψ^{-2} and δ^{-2} are averages of the squared heading and rudder deviations, respectively. Typical values of $\mu = 0.014$ deg^{-2} and $\rho = 0.1$ have been suggested for a tanker. Thus Aström and his co-workers have used a performance criterion of

$$v = \frac{1}{T}\int_0^{\cdot} [\psi^2(t) + \rho\delta^2(t)]\, dt$$

In the adaptive autopilot studies of Källström *et al.*[29, 30] a dual-mode operation was implemented covering the phases of steady course-keeping and turning. For course-keeping, a self-tuning regulator minimizing the above loss function was used. The parameters of the model corresponding to eqn. (6.20) were estimated using least-squares identification,[31] while a simplified control algorithm based on minimized predicted heading errors was found to be satisfactory. Such a regulator cannot, however, deal with large heading changes during maneouvres because of the non-linear behaviour of the ship. The turning controller was designed to be a high-gain regulator with constant parameters for each of four phases comprising initiation, steady-state turning rate, stopping and fine adjustments, considered separately. Two autopilots were constructed, one using measurements of forward speed and heading only, while the other used Kalman filtering on measurements of speed, heading, yaw rate, rudder angle, fore and aft sway velocities. In addition, velocity scheduling was used to update the parameters of the Kalman filter, the self-tuner and the turning regulator according to forward speed. Although this is not necessary for the self-tuner, it was claimed to give faster adaptation. In simulation studies for large tankers, improvements in drag of 0·3% for the simpler autopilot over a manually adjusted PID controller were reported. Using the Kalman filter autopilot, improvements of 0·6 to 1·4% were reported with biggest improvements being at full draught and high wind speeds. In sea trials on three large tankers covering about 100 hours, similar reductions in drag of 0 to 2% were recorded over a wide range of operating conditions. A comparison between PID control and the simpler autopilot is given in Fig. 6.11. The turning regulator was tested under both small and large course

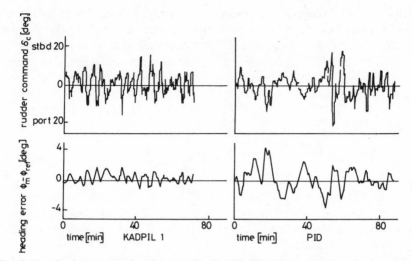

Fig. 6.11 *Straight course-keeping on the 'Sea Swift' using a simple adaptive autopilot KADPIL 1 and a well-tuned PID controller. These conditions gave 2.7% less drag with the adaptive autopilot*
(From Källström, C. G., Aström, K. J., Thorell, N. E., Eriksson, J., and Sten, L., 'Adaptive Autopilot for Large Tankers', IFAC World Congress, Helsinki, 1978)

changes and gave good performance, an example of which is shown in Fig. 6.12.

Experimental investigation into the form and parameters of a suitable performance criterion have been carried out on a radio-controlled model of a container ship.[32] These results supported the form of index adopted by Aström and his co-workers.

Another approach to adaptive ship steering is based on the non-linear simplified dynamic of eqn. (6.15) involving the Bech polynomial $H(\dot{\psi})$ characteristic. A model-reference adaptive autopilot has been investigated via simulation and in sea trials on the *Capella* (unfortunately a stable ship).[33, 34] In this method the parameters in eqn. (6.15) were taken to be constant and known except for the $H(\dot{\psi})$ characteristic assumed to consist of $a\dot{\psi}^3 + b\dot{\psi}$. Parameter a was assumed to be constant and b to be time-varying to account for changes in operating conditions, particularly caused by water depth variations. An integral error square criterion was used to give adaptive control of a rate feedback term to give suitable response. The error was the difference between the ship and model time-responses. The adaptive loop was designed using both sensitivity functions, corresponding to continuous hill-climbing, and the Liapunov method. In sea-trials it was observed that in the presence of noise a low-pass filter was necessary in the Liapunov autopilot, and also that high-frequency rudder movements could be avoided using low-pass filtering. Also, in the presence of noise the adapted parameter drifted to an incorrect value in the absence of course changes.

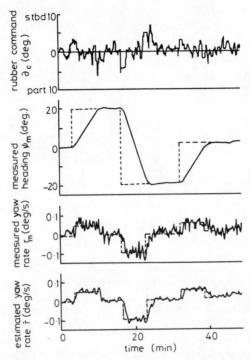

Fig. 6.12 *Turning regulator experiment on the 'Sea Stratus' at a speed of 13 knots under ballast*
(From Källström, C. G., Aström, K. J., Thorell, N. E., Eriksson, J., and Sten, L., 'Adaptive Autopilot for Large Tankers', IFAC World Congress, Helsinki, 1978)

6.5 Conclusions

Although the basic equations of motion are similar for aircraft and ships the practical implications are very different. The concept of hydrodynamic derivatives to characterize fluid forces on a body has been successfully employed in aircraft design and in submarine and fast surface ship analysis. For large commercial ships, particularly of the supertanker type, the use of linear dynamic models in autopilot design becomes questionable. For large manoeuvres, consideration of the non-linear dynamic components of motion becomes necessary and has led to designs based on model-reference adaptive schemes and switching-mode PID regulators. Clearly, the difficult question of identification of non-linear dynamics is an important aspect for future work. Some preliminary work on self-tuning regulation of non-linear ship models has shown promising results.[35]

For course-keeping under small disturbances, linear identification and control have been well investigated for a range of commercial vessels. These studies have shown that model orders of 2 or 3 are sufficient to give good identification and that system estimation techniques such as maximum-likelihood are at least as successful as the more classic manoeuvring tests developed over the years for acceptance sea trials.

Improvements in drag performance caused by modern identification and control methods are apparently small, but are significant in reducing fuel costs in transportation. The increasing importance of minimizing such costs may make it both desirable and necessary to improve the level of model sophistication and accuracy towards that obtained for aircraft and missile guidance. If this is the case then advances in techniques of non-linear modelling and estimation will become increasingly important.

6.6 References

1 ABKOWITZ, M. A.: 'Stability and Motion Control of Ocean Vehicles' (MIT Press, 1969)
2 BISHOP, R. E. D., and PARKINSON, A. G.: 'Choice of origin for body axes attached to a rigid vehicle', *J. Mech. Eng. Sci.*, 1969, **11**(6), p. 551–554
3 MANDEL, P.: 'Ship manoeuvring and control', *in* COMSTOCK, J. P. (Ed.) 'Principles of Naval Architecture', (SNAME, New York, 1967)
4 BECH, M., and SMITT, L. W.: 'Analogue simulation of ship manoeuvring', Hydro-og. Aerodynamisk Laboratorium, Lyngby, Denmark, Sept. 1969, Report No. Hy-14
5 NOMOTO, K.: 'On the steering qualities of ships', *Int. Shipbuilding Progress*, 1957, **4**(35)
6. SMITT, L. W.: 'The reversed spiral test—a note on Bech's spiral test and some unexpected results of its application to coasters', 1967, hyA Report Hy-10
7 CLARKE, D.: 'A new non-linear equation for ship manoeuvring', R. Council/British Ship Research, pp. 1–24.
8. BISHOP, R. E. D., BURCHER, R. K., and PRICE, W. G.: 'The uses of functional analysis in ship dynamics', *Proc. R. Soc. London A.*, 1973, **332**, pp. 23–35
9 BISHOP, R. E. D., BURCHER, R. K., and PRICE, W. G.: 'Application of functional analysis to oscillatory ship model testing', *Proc. R. Soc. Lond. A.*, 1973, **332**, pp. 37–49
10 VAN OORTMERSSEN, G.: 'Influence of water depth on the manoeuvring characteristics of ships', NSMB Symp. on 'Ship Handling', 1973
11 FUJINO, M.: 'The effect of frequency dependence of the stability derivatives on manoeuvring motion', *Int. Shipbuilding Progress*, 1975
12 RAINEY, R. C. T., and MASON, P.: 'Low speed manoeuvring characteristics of large vessels in shallow water', IMC Symp. 'Dynamic Analysis of Vehicle Ride and Manoeuvring Characteristics', London, Nov., 1978, pp. 65–78
13 GERTLER, M.: 'The DTMB planar-motion-mechanism system', Symp. on 'Towing Tank Facilities, Instrumentation and Measuring Technique', 1959
14 GOODMAN, A.: 'Experimental techniques and methods of analysis used in submerged body research', Reprint—3rd Symp. on Naval Hydrodynamics High Performance Ships, 1960, Office of Naval Research, Dept. of the Navy, Washington, D.C., Sept. 19–22.
15 VAN LEEUWEN, G.: 'Some notes on the discrepancies between the lateral motions of oscillation tests and full scale manoeuvres', Planar Mot. Mech. P.G. course, Haslar A.E.W., 21–23 April, 1960, notes
16 STROM TEJSEN, J., and CHISLETT, M. S.: 'A model testing technique and method of analysis for the prediction of steering and manoeuvring qualities of surface vessels', Hydro-og Aerodynamisk Laboratorium, 6th Symp. on Naval Hydrodynamics, Washington, D.C., Aug. 1966, Rep. No. Hy-7

17 NOMOTO, K.: 'Analysis of Kempf's standard manoeuvre test and proposed steering quality indices', Proc. 1st Symp. on 'Ship Manoeuvrability', 1960, Washington, D.C. (DTMB Report No. 1461)

18 WINDETT, G. P., and FLOWER, J. O.: 'Measurement of ship roll dynamics by pseudorandom binary sequence techniques', *Journal Naval Science*, 2(1), pp. 41–48

19 KAPLAN, P., SARGENT, T. P., and GOODMAN, T. R.: 'The application of system identification to dynamics of naval craft', 9th Symp. on 'Naval Hydrodynamics', 1972, Paris

20 ASTRÖM, K. J., and KÄLLSTRÖM, C. G.: 'Identification and modelling of ship dynamics', Report 7202, Dept. of Auto. Cont., Lund Inst. of Tech., 1972

21 ASTRÖM, K. J., NORRBIN, N. H., KÄLLSTRÖM, C. G., and BYSTRÖM, L.: 'The identification of linear ship steering dynamics using maximum likelihood parameter estimation', SSPA Report 1920-1, 1974

22 ASTRÖM, K. J., and KÄLLSTRÖM, C. G.: 'Identification of ship steering dynamics', *Automatica*, 1976, **12**, pp. 9–22

23 TIANO, A., PIATELLI, M., and LECCISI, D.: 'Computer control of ships steering', Report Laboratorio per l'Automazione Navall, Genoa, 1973

24 AKAIKE, H.: 'Use of an information theoretic quantity for statistical model identification', Proc. 5th Hawaii Int. Conf. on 'Systems Science', 1972, Western Periodicals Co.

25 TIANO, A.: 'Identification and control of the ship steering process', in 'Ship Operation Automation', Pitkin, Roche and Williams, (Eds.) 1976, North-Holland Pub. Co.

26. ASTRÖM, K. J.: 'Some aspects of the control of large tankers',

27 BROOM, D. R., and LAMBERT, T. H.: 'The effects of variations of autopilot parameters on the yaw motion response of ships under sea excitation', IMC Conf. on 'Dynamic Analysis Vehicle Ride and Manoeuvring Characteristics', London, Nov. 1978, pp. 21–29

28 NORRBIN, N. H.: 'On the added resistance due to steering on a straight course', 13th Int. Towing Tank Conf., 1972, Berlin/Hamburg

29 KÄLLSTRÖM, C. G., ASTRÖM, K. J., THORELL, N. E., ERIKSSON, J., and STEN, L.: 'Adaptive autopilots for large tankers', IFAC World Congress, 1978, Helsinki, pp. 477–484

30 KÄLLSTRÖM, C. G., ASTRÖM, K. J., THORELL, N. E., ERIKSSON, J., and STEN, L.: 'Adaptive autopilots for steering of large tankers', Kockums Automation AB Report LUTFD2/(TFRT-3145), 1977

31 ASTRÖM, K. J., and WITTENMARK, B.: 'On self-tuning regulators', *Automatica*, 1973, **9**, p. 185

32 BROOME, D. R., and LAMBERT, T. H.: 'An optimising function for adaptive ships autopilots', 5th 'Ship Control Systems' Symp., Oct., 1978, Maryland, USA

33 HONDERD, G., and WINKELMAN, J. E. W.: 'An adaptive autopilot for ships', 3rd 'Ship Control Systems' Symp., 1972, MOD, Bath, England

34 VAN AMERONGEN, J., and UDINK TEN CATE, A. J.: 'Model reference adaptive autopilots for ships', *Automatica*, 1975, **11**, pp. 441–449

35 MORT, N., and LINKENS, D. A.: 'Self-tuning controllers for surface ship manoeuvring', IMC Symp. 'Dynamic Analysis of Vehicle Ride and Manoeuvring Characteristics', London, Nov., 1978, pp. 53–63

Chapter 7

Biological systems

D. A. Linkens

7.1 Introduction

Biological systems are concerned with living organisms which are dynamic by nature because of their changing behaviour both in time (the life cycle) and in space (movement and transportation). Some organisms, such as unicellular plants and animals, are relatively simple; others, such as the higher mammals, far exceed in complexity the most sophisticated technological processes described in other chapters of this book. In this chapter the emphasis is on the complex interrelating dynamic mechanisms of the mammalian body. Paradoxically, more is known about the mammalian control mechanisms than about those of simpler animals. This is largely due to the much greater amount of research which has been devoted to mammalian physiology.

It has been known for many years that the biological world contains many feedback mechanisms and dynamic structures, stemming ultimately from Greek philosophers such as Hippocrates (460–377 B.C.). A century ago physiologists such as Claude Bernard had been impressed by the dynamic stability of physiological parameters such as body temperature, and observed that for an organism to function optimally its component cells must be surrounded by a medium of closely regulated composition. Such a view has been substantiated, so that it is clear that higher vertebrates contain a multiplicity of complex regulatory dynamic systems and, in consequence, the interior environment is controlled to within very fine limits. In the words of W. B. Cannon, in 1929, 'The co-ordinated physiological processes which maintain most of the steady states in the organism are so complex and so peculiar to living beings—involving as they may, the brain and nerves, the heart, lung, kidneys and spleen, all working co-operatively—that I have suggested a special designation for these states, *Homeo stasis*'. It is an elucidation of some of the subsystems that make up this remarkable homeostasis that is the content of this chapter.

The opening section describes the classical work of Stark who used system identification techniques employing forcing functions of steps, sine-waves and random disturbances to investigate the eye pupil regulatory mechanism. These elegant experiments showed how systems theory can be applied to living organisms, but quickly came across two almost universal difficulties when investigating biological dynamics. The more detailed the investigation is, the more non-linear the dynamics become, and the greater are the interacting influences from other mechanisms, i.e. system boundaries cannot be isolated. The next section shows how similar identification techniques are currently being used to study the regulation of breathing with respect to human performance. In a realm where little is known about the dynamics of the process, linear methods at present show promise of shedding light on the underlying physiology. Presumably the problems of non-linearity and interaction will again arise in due course.

While the eye can be stimulated optically, and the breathing performance via workload regimes, the human digestive tract is an internal organ with little obvious access for testing stimuli. Over the past decade, however, much data have been collected regarding electrical potentials in the gut which are considered to be the initiators of mechanical contractions responsible for food transport. In this case the electrical behaviour is that of a continuously oscillating system, which allows data to be collected even when there are no experimental forcing functions present. Electrical stimuli have, however, been successfully used and given valuable information about the electrical structure and its correlated mechanical transport system. The dynamic model is essentially non-linear, but interactions from the central nervous system are secondary in nature, although important for adequate functioning of the gut.

The remaining sections are shorter but further illustrate the above ideas. Very detailed mathematical modelling has been applied to nerve impulse generation and conduction, coupled to extensive physiological measurements. The relationship between measurement, mathematics and ionic structure is seen in this work which has motivated many physiological studies not only in nerve tissue but also in muscle tissue. The glucoregulatory system is a major example of homeostasis in the body, defects in which result in such diseases as diabetes. It is a system which interacts with many organs in the body, but modelling studies even with relatively small systems boundaries soon become highly complex and non-linear. One excursion is made into the modelling of much simpler systems, being that of yeast cell growth. In this case, considerable information is known about the biochemical pathways, and the dynamics can be associated quite clearly with particular reactions.

The following sections are introductory in nature, but for some of the systems described further information is contained in a companion book on Biological Systems, Modelling and Control in this Series.[1] For the remaining systems the references given will lead the interested reader into further details.

7.2 Reflex regulation of pupil area

The automatic regulation of pupil area has long been recognized as an example of feedback control, and has been intensively studied using control identification techniques in work pioneered by L. Stark. The pupil has two major functions: regulation of the amount of light reaching the retina, and assistance in maintaining image sharpness during near-vision. This section is concerned only with the former of these two functions.

In man the pupil is circular, and its diameter is controlled by two antagonistic muscles. The dilator muscle is connected to the sympathetic nerve system, and the sphincter muscle is controlled via the parasympathetic nerves. The sphincter muscle is of major importance in pupillary regulation. Variations in pupil area cause changes in illumination of the retina which produce changes in the feedback signals transmitted via the optic nerve to the brain stem. The neural pathways serving the two eyes are cross-coupled, so that illuminating one eye causes constriction of both eyes to occur. This does not, however, affect the behaviour of the single-loop feedback dynamics which are now described.

A simplified block diagram of the pupil control system is shown in Fig. 7.1. The retina is sensitive to the light flux F with a logarithmic law. The retina also exhibits slow photochemical dark-adaption and relatively fast light-adaption, giving rise to non-linear behaviour. The sphincter muscle is itself a complex subsystem involving a number of dynamic processes, whose transfer function was experimentally determined by Stark.[2] Before giving details of the transfer functions, some brief comments on the methods of applying stimuli to the eye will be made.

A suitable forcing function in this case is a pencil of light shone directly at the eye. If the light illuminates the whole of the pupil, then the intact closed-loop pupil system is being stimulated, in which case the well-known problems of identification of internal dynamics within a closed loop are encountered. There is, however, a simple way to experimentally open the loop while leaving

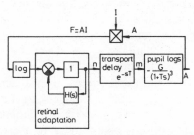

Fig. 7.1 *Simplified block diagram of the closed-loop responsible for the pupillary light-reflex* (From McFarland, D. J. (1971) 'Feedback Mechanisms in Animal Behaviour', fig. 4.14, Academic Press)

the organ intact. This entails illuminating only the central part of the pupil so that light entering the pupil is unaffected by movements of the iris. This method was adopted in initial experiments by Stark and Sherman,[3] in determining sphincter muscle dynamics. A further simple technique which can be used to artificially increase the loop gain of the pupil mechanism is to shine a narrow beam of light onto the border of the iris and the pupil. It has been known since 1944 that this technique can induce pupillary oscillations.[4] Another method used by Stark[5] was to artificially close the loop via the pupillometer which measures pupil diameter. By processing this signal and using it to control the stimulus light intensity, different controller gains and functions could be synthesized and used to probe the internal dynamics of the pupil reflex loop.

Because of the non-linear nature of the sphincter muscle dynamics, classic frequency analysis is not truly applicable, but using small sinusoidal inputs Stark obtained a linear approximation. Under open-loop stimulus conditions the pupil response was approximately sinusoidal, but also exhibited high frequency noise and low frequency drifting. The amplitude and phase response curves obtained in these experiments are shown in Fig. 7.2. The amplitude response was low-pass in character with a high frequency asymptote of about -18 dB/octave indicating a third-order system. The phase response showed much greater lag than for a third-order system, suggesting the presence of a pure time delay in the dynamics. Thus, Stark estimated the open-loop transfer function to be

$$G(s) = \frac{0 \cdot 16 e^{-0 \cdot 18s}}{(1 + 0 \cdot 1s)^3}$$

From this transfer function, the low frequency closed-loop gain was calculated to be $0 \cdot 14$ which compared favourably with an experimentally determined closed-loop gain of $0 \cdot 15$.[2] Thus, despite the known non-linearity in the pupil system, small-signal frequency analysis yielded acceptable results and gave an estimate for the time delay of the same order as that obtained from transient analysis.

Transient analysis is normally performed using step or pulse inputs to the system, in this case using changes in illumination level. Clynes[6] used four types of stimulus being light-pulses and light-steps (i.e. increases in illumination), and dark-pulses and dark-steps (i.e. decreases in illumination). The responses to these stimuli are shown in Fig. 7.3 which clearly demonstrate the non-linear property of the pupil control mechanism. These dynamic responses can be modelled using three components: (1) a time-delay of about 250 ms; (2) a proportional response, sensitive to illumination changes in both directions; (3) a rate response, sensitive only to increases in illumination. Thus, Clynes suggested the following transfer function from these experiments

$$H(s) = \frac{b e^{-T_5 s}}{1 + T_4 s} + \Omega \frac{a s e^{-T_3 s}}{(1 + T_1 s)(1 + T_2 s)}$$

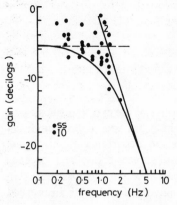 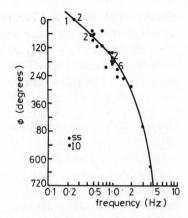

Fig. 7.2 *Frequency responses of pupillary mechanism for both high gain instability experiments (numbers) and driven response experiments (filled circles)* (From Stark, L. (1968) 'Neurological Control Systems. Studies in Bioengineering', Plenum Press)

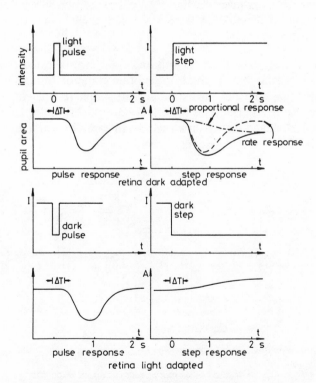

Fig. 7.3 *Transient responses of the pupil to pulse and step inputs* (From Clynes (1961) *Ann. N.Y. Acad. Sci.*, **93**, 846–969)

where Ω is an operator given by

$$\Omega = 0 \qquad \text{for} \qquad \frac{dx}{dt} < 0$$

$$\Omega = 1 \qquad \text{for} \qquad \frac{dx}{dt} \geq 0$$

The first term represents the proportional response of the eye, and the second term represents the non-linear rate response which has been called 'unidirectional rate sensitivity' (URS). URS is a common occurrence in biological systems, particularly where release of chemical substances is involved, because the release rate is much faster than the decay rate. It can also occur in muscular contractions and both effects may be involved in the URS observed in the pupillary light reflex.

The third type of analysis commonly used in dynamic modelling is that due to noise in the system, and this has also been investigated by Stark and his co-workers for the pupil mechanism.[7] From a range of experiments it was observed that pupil noise level was Gaussian in nature and varied with the level of light stimulus. It was found that the r.m.s. value of pupil noise area signal was a linear function of average pupil area, suggesting that noise affects the system in a multiplicative manner. Autocorrelation functions of the pupil were essentially the same under high and low levels of illumination, while cross-correlation of noise from the left and right eye indicated that noise is introduced or generated at a point in the brain common to both eyes. From these and other experiments Stark modified the simple retinal adaptation part of Fig. 7.1 to that of Fig. 7.4. This illustrates how quickly biological modelling can develop from analysable linear structures to complex non-linear dynamics beyond analysis via normal systems theory.

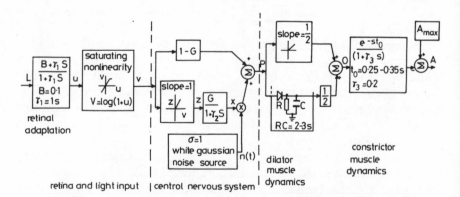

Fig. 7.4   *Non-linear model of pupil reflex model accounting for pupil noise phenomena* (From Stark, L. (1968) 'Neurological Control Systems. Studies in Bioengineering', Plenum Press)

This section has summarized an extensive set of experiments based on systems techniques of frequency, transient and stochastic analysis performed to elucidate the dynamics of the pupillary light reflex. While it is now evident that such a system is highly non-linear and complex, it should also be realized that it is not an isolated system but is interrelated to other mechanisms such as focal length adjustment of the eye lens, and three-dimensional rotation of the eye to direct it towards the centre of interest. Add to this the fact that eye stimuli to the brain are linked with hearing and balance signals and one begins to realize the enormous and wonderful nature of interacting dynamic control of the human body. Some of these further loops in the audio-visual control mechanism are described and modelled in Chapter 3 of Reference 1.

7.3 Performance testing of human respiratory regulation

The previous section described dynamic modelling studies performed a decade ago on the eye using a range of conventional system forcing functions. In this section current research work using similar techniques on an analysis of the human respiratory control system is described.

The first dynamic model of respiration regulation was due to Grodins in 1954[8] and considered the ventilation response to step changes in inspired CO_2. Many other models have followed which have included more descriptions of the underlying physiology and attempted to explain a larger number of experimental observations, e.g. Grodins *et al.*[9] The tendency for biological modelling to increase in its complexity is again illustrated in this evolution of respiratory models. These different model structures reflect the controversies concerning the physiology of human respiratory regulation, and thus represent alternative hypotheses, all competing to describe the same biological process. The development and use of modelling techniques which will discriminate between such hypotheses is clearly valuable in forwarding research in this area.

It is important to distinguish between the controlled system (the lung) and the controller (the respiratory centre). We shall consider here the problem of identifying the controller dynamics, which has as an input the alveolar $CO_2 - O_2$ gas concentrations and an output of alveolar ventilation. These variables are not directly accessible for measurement, but estimates of them can be made via end-tidal gas measurement and tidal volume with breath cycle timing instrumentation.

In order to identify the controller dynamics it is necessary to artificially open the respiratory regulation loop. To do this, the end-tidal gas concentrations are forced to follow particular wave-forms, independent of ventilation. Conversely, if the controlled system is to be identified, then the ventilation is forced to follow a given wave-form independent of the arterial gas tensions. Experimentally, these concepts can be implemented using dynamic end-tidal forcing techniques and voluntary ventilation techniques. Dynamic end-tidal forcing utilizes prediction and correction of inspired concentrations to achieve the desired end-tidal wave-forms.[10] Voluntary ventilation utilizes audio and visual cue feedback to the subject, to enable him to force his ventilation to follow a specific wave-form on a breath-to-breath basis.[11] This feedback consists of an inhaled volume displayed on an oscilloscope, and breath sounds generated on a computer-controlled breath-sound simulator.

The dynamics of the respiratory controller have been estimated by a number of workers using different forcing functions, resulting in models with first-order dynamics only,[12] and models with time-delay and second-order dynamics.[13] The major debate is whether the controller contains a 'fast' component (associated with neural behaviour?) as well as a 'slow' component (associated with humoral behaviour?). The use of steps, sine-waves and PRBS has helped to distinguish between these models, and it appears to be generally accepted now that there is indeed a 'fast' component, particularly during exercise conditions. A model which is both structurally identifiable and isomorphic to an understanding of respiratory physiology has been proposed by Swanson and Bellville.[14] The model is given by

$$\dot{x}_1 + \alpha_1 x_1 = \alpha_1 g_1 u(t - t_d) - k \tag{7.1}$$

$$\alpha_1 = mu(t - t_d) + b \tag{7.2}$$

$$\dot{x}_2 + (1/\tau_2)x_2 = (1/\tau_2)g_2 u(t - t_d) - k \tag{7.3}$$

$$y = x_1 + x_2 \tag{7.4}$$

Equations (7.1) and (7.2) are related to the central chemoreceptor response characteristics, while (7.3) is related to the peripheral chemoreceptor response characteristics. The model input $u(t)$ is the end-tidal CO_2, and the model output $y(t)$ is the ventilation. The bias term is analogous to the so-called 'intercept' in the steady-state response. The delay time, t_d, is related to the circulation time. The central and peripheral chemoreceptor steady-state gains are given by g_1 and g_2, respectively. α_1 is a function of the delayed input $u(t - t_d)$, which is consistent with cerebral blood flow coupled to arterial blood CO_2. The parameter m is related to the slope of the linearized curve relating cerebral blood flow to arterial CO_2, while b is the intercept. Finally, parameter τ_2 is related to the speed of response of the peripheral chemoreceptors. Estimation of the parameters in eqns. (7.1) to (7.4) and testing its validity are two important areas of interest.

The conventional systems forcing functions of steps, sine-waves, ramps, and PRBS have all been used to identify the respiratory controller. Step-response testing tends not to distinguish relatively fine frequency response characteristics, and as a result is not the best forcing function to discriminate between the Casaburi and Fujihara models. Whipp *et al.*[15] were, however, able to discern a 'fast' component using step inputs, particularly for high muscular work loads. It should be noted that these experiments were also conducted using staircase, ramp and sinusoid (periods of 10·0, 4·0, 2·0 and 0·7 minutes) inputs, and hence a comprehensive set of results was obtained enabling particular features of the system dynamics to be detected. Bennett *et al.*[16] used PRBS testing to distinguish between the two hypothesized models. This was attempted to allow investigation over a wide frequency range and avoid excessive work rates. The subject pedalled on an ergometer with the work rate being switched between 25 W and 100 W with a PRBS of length 63 and clock period 5 seconds. Five subjects were tested, and the average of 30 cross-correlations determined for each subject. A Nelder-Mead search algorithm was used to fit each model to the cross-correlation impulse response of the subject. It was found that the Fujihara 'fast' component model gave a significantly better fit to the data. Reducing the bandwidth by a factor of 4 failed to give identification of the 'fast' component, suggesting that the sinusoidal testing range in Ref. 12 was insufficient to fully excite the respiratory controller dynamics.

In estimating the parameters in eqns. (7.1) to (7.4) Swanson[17] used both periodic switching (square wave) and non-periodic switching functions. He found that the variance in the parameters was improved by about 15% when using a non-periodic switching strategy. He also found that there were optimal frequencies for estimating each physiological parameter (see Fig. 7.5). The cerebral blood flow parameters (m, b) and steady-state intercept (k) were

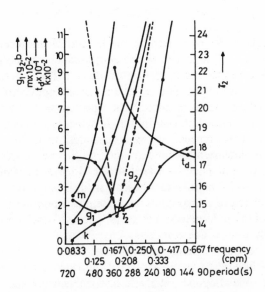

Fig. 7.5 *First-order variances for estimation errors in parameters of the breathing regulation model of eqns. (7.1) to (7.4)*
(From Swanson, G. D. (1977) *Proc. IEEE*, pp. 735–740)

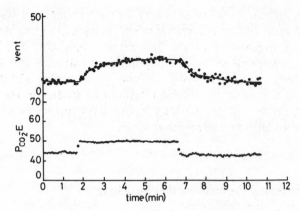

Fig. 7.6   *Experimental and model step response to changes in end-tidal CO_2*
(From Swanson, G. D. (1977), Am. J. Physiol., 233, pp. 66–72)

best estimated using a low-frequency forcing function. The central and peripheral chemoreceptor parameters (g_1, g_2, τ_2) were estimated best at middle frequencies, while the circulation delay parameter (t_d) was best estimated at high frequencies. An example of model response and human performance to a square wave perturbation in end-tidal CO_2 is shown in Fig. 7.6 for optimized parameters. The problem of selecting the best input signal to discriminate between hypothesized controller dynamics has been pursued further in a novel way by Swanson.[18] The input signal was chosen to comprise a summed set of 7 sinusoids whose amplitudes and frequencies were optimized to give maximum difference between outputs of the models which were to be discriminated. The basis of the choice of an input comprising 7 summed sinusoids is the minimum variance characteristic of Fig. 7.5. The generation and application of variable amplitude sinusoids to a physiological system is seldom convenient. Thus, Swanson replaced the optimal sinusoid wave-form with a bang-bang forcing function having an equivalent autocorrelation function. Both of these types of wave-form were successful in distinguishing between the Casaburi and Fujihara models.

Using a similar model to eqns. (7.1) and (7.4) Wiberg *et al.*[19] investigated parameter estimation algorithms based on least squares and maximum likelihood. The forcing function was a single on-and-off step transition (see Fig. 7.6). Simple least squares gives a biased estimate if the process residuals are correlated. The autocovariance and power spectral density of the residuals were plotted to check for whiteness. It was found that least squares and maximum likelihood gave similar results for normal breathing conditions, whereas a bias occurred in the least-squares estimates under abnormal conditions. They concluded that in drug studies, where changes in parameters and small parameter variances are important, the method of maximum likelihood is preferable for system identification.

From this section it can be seen that all the conventional systems testing and identification techniques are currently being utilized to determine the structure and parameters of the respiratory controller. The debate will continue in attempting to localize physiological pathways (see Ref. 20), and modelling studies are undoubtedly assisting in the testing and destroying of hypotheses in this area. In a similar fashion to studies in the pupil reflex regulation it surely will not be long before the question of non-linear dynamics enter an already controversial scene.

7.4 Gastro-intestinal electrical rhythm modelling

In the past 15 years a steadily increasing amount of data has become available about the spontaneous electrical rhythms which exist in many parts of the digestive tract in mammals. Research in this area has lagged behind that of the heart ECG, partly because it is not possible at present to isolate specialized areas of smooth muscle gut tissue corresponding to cardiac 'pacemaker' tissue. Modelling studies have, however, been keeping pace with data collection and physiological studies, and are described briefly in this section.

The electrical potential rhythms comprise two components: 'slow waves' and 'spike activity'. They occur in all parts of the digestive tract below the middle of the stomach. A schematic diagram of the tract together with typical 'slow wave' frequencies recorded in man is shown in Fig. 7.7. The frequency and wave-shape of the 'slow waves' vary considerably between the organ and the species being studied. For example, the canine stomach has narrow pulse-like waves of about 0·08 Hz, while the human stomach produces square-like waves of about 0·05 Hz. The human small intestine has a falling gradient, but with a constant frequency of about 0·2 Hz in the first part of the duodenum, falling to about 0·17 Hz in the ileum.[21] The constant frequency in the duodenum is commonly called a 'plateau', and plays an important part in modelling and analytical investigations. The wave-shape here is nearly sinusoidal when using recordings from intact organs. In the human large intestine a more complex pattern exists, comprising periods of electrical silence interspersed with two rhythms of about 0·05 Hz and 0·12 Hz which are not harmonically related.[22] Large-intestinal wave-forms are nearly sinusoidal and show considerable variations in amplitude. Amplitude variations also occur in human small-intestinal rhythms and recent spectral analysis of these variations has shown rhythmic behaviour which is relevant to modelling studies.[23]

Synchronized with the 'slow waves' are action potentials, commonly called 'spike activity', which occur whenever there are mechanical contractions in the tract. The nature of the 'spike activity' also varies considerably between organ and species. A further phenomenon is a travelling-wave of localized

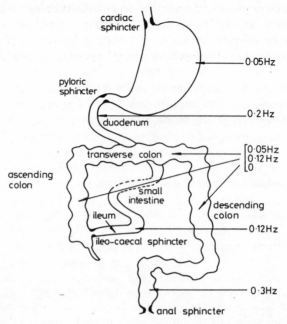

Fig. 7.7   *Digestive tract schematic diagram showing typical 'slow-wave' frequencies*

'spike activity', referred to as a 'migrating complex', which occurs in the stomach and small intestine.[24, 25] 'Slow waves' are considered to act as a co-ordinator of the 'spike activity' with their related mechanical muscular contractions. Neural and hormonal effects in intact organs appear to be small, in that changes in slow-wave frequencies of greater than 10% are rare even for large stimulation by drugs. Neural influences appear to be modulating effects on the rhythms produced within the muscle tissue itself, evidenced by the changes in electrical activity caused by cutting nerve trunk supplies to the stomach. Further details about gut rhythm knowledge can be found in a review paper by Duthie.[26]

Earliest attempts at 'slow-wave' modelling suggested the existence of localized 'pacemaker' tissue with propagation along the organ, analogous to cardiac muscle. The fact that small pieces of excised tissue from most parts of the gut exhibit spontaneous oscillations, plus the inability to identify a specialized pacemaker tissue structure in the gut, have mitigated against this model. An alternative model which has now found wider acceptance and which has been shown to be capable of reproducing the major physiological phenomena was first hypothesized by Nelsen and Becker.[27] The model comprises a set of mutually coupled non-linear oscillators. The small and large intestines have a tubular structure, and show almost complete electrical co-ordination around the circumference of the organ. Hence, one-dimensional chains of coupled oscillators have been investigated for intestinal modelling.

In contrast, large phase shifts and frequency differences occur along both axes of the stomach, so that two-dimensional arrays of coupled oscillators have been necessary for gastric modelling. Coupling between oscillators has been considered to be linear, comprising parallel resistive, capacitive and inductive networks.

A number of different system dynamics for the component oscillators have been investigated for the coupled models. The dynamics have always been a non-linear form because of the limit-cycle nature of gut slow waves. This at once precludes linear systems analysis for examination of such models, but at the end of this section progress in analytical treatment of coupled non-linear oscillators will be described briefly. The simplest, and most commonly used, oscillator dynamic is that of van der Pol's equation. Including its coupling terms this can be written as follows for the nth oscillator and a one-dimensional chain

$$\ddot{x}_n - \epsilon_n(a_n^2 - x_n^2)\dot{x}_n + \omega_n^2 x_n - Lx_{n+1} - R\dot{x}_{n+1}$$
$$- C\ddot{x}_{n+1} - Lx_{n-1} - R\dot{x}_{n-1} - C\ddot{x}_{n-1} = 0 \qquad (7.5)$$

The equivalent circuit form of this equation is shown in Fig. 7.8 which identifies the RLC coupling parameters. Sarna *et al.*[28] reported a 16-oscillator analogue simulation for the canine small intestine using the Fitzhugh generalization for van der Pol's equation given by

$$\dot{x} = \alpha(ey + fx + gx^2 + hx^3)$$
$$\dot{y} = -\frac{1}{\alpha}(by + \omega^2 x + cx^2 + dx^3 - a) \qquad (7.6)$$

To represent the whole of the small intestine with 16 oscillators is clearly a gross simplification, and Robertson-Dunn and Linkens[29] increased this to 100 oscillators in a digital simulation of the human small intestine. In these simulations the entrainment of oscillators at a common frequency when there was a gradient of intrinsic frequencies was a necessary feature. This is il-

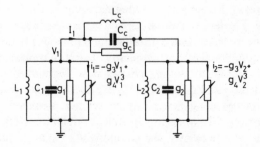

Fig. 7.8 *Equivalent circuit representation of two van der Pol oscillators coupled with a parallel RLC network*

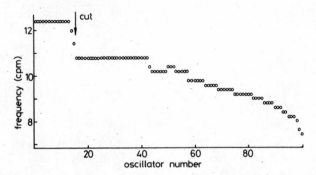

Fig. 7.9   *One-hundred-oscillator small-intestinal digital simulation showing a secondary plateau formed below an 'incision'*

lustrated in Fig. 7.9 which also shows the effect of an artificial cut in the model which reproduced the secondary plateau below the incision observed in animal experiments. Travelling-wave patterns of frequency perturbations have been studied for this model, and it has been found that resistively coupled chains give stable behaviour with transients being self-cancelling. These travelling waves did not appear to correspond in nature to the 'migrating complex' of the gut, giving support to the view that this phenomenon is caused by the extrinsic nerve supply to the tract.[30]

A reactively coupled oscillator model gives rise to multiple modes and multimodes which appear to be relevant to large intestinal electrical rhythms. The large intestine is not a well-coordinated organ electrically, and it is considered that this is caused by relatively weak intercoupling between oscillating units.[31] A hypothesized model for the human large intestine comprises a reactively coupled chain of modified van der Pol oscillators with dynamics given by

$$\ddot{x} + \epsilon(b - cx^2 + dx^4)\dot{x} + \omega^2 x = 0 \tag{7.7}$$

Suitable choice of b, c and d in (7.7) produces a stable equilibrium point of zero (i.e. electrical silence), and one stable limit cycle. Coupling together two or more oscillators of this type using either capacitive or inductive terms gives two stable modes whose frequencies depend on coupling strength. Computer simulations occasionally showed another mode in which both frequencies were present simultaneously. Subsequently, this multimode behaviour has been observed physiologically[32] and predicted analytically.[33]

A major disadvantage of using van der Pol dynamics as described above is that the individual parameters cannot be related directly to physiological biochemical processes. In an attempt to relate modelling studies and physiological experiments more closely, the component oscillators have been simulated using modified versions of the Hodgkin-Huxley equations. These equations were developed to model nerve impulses in the squid axon and are further described in Section 7.5. Using a slightly modified set of Hodgkin-Huxley equations which give limit-cycle oscillations it has been shown that a coupled oscillator model of this type is capable of frequency entrainment requisite in intestinal modelling.[34] An example of this is shown in Fig. 7.10 for two resistively coupled oscillators with different uncoupled frequencies, showing the phase shift that accompanies entrainment. The numerical problems in digital simulation of a large number of such coupled oscillators are considerable, and prohibitive computing time is involved. A small-intestinal model comprising 30 Hodgkin-Huxley type oscillators has, however, been simulated and shown to give travelling waves of instantaneous frequency similar to those for a van der Pol model and also a 'plateau' entrainment region at the top of the model.[35]

Because of the long computing times involved in the above models an electronic version of a coupled Hodgkin-Huxley system has been developed.[36] This enables experiments to be performed with hands-on adjustment of parameters. A wide range of oscillator waveforms is available using this hardware, some of which are illustrated in Fig. 7.11. These types of wave-shapes are all of relevance in smooth muscle gut simulations, since wave-forms are quite variable between species and organ in the tract. Electronic versions of van der Pol dynamics have also been constructed and used in comparative studies with the more complex Hodgkin-Huxley dynamics. A 64-oscillator small-intestinal model has also been investigated in which the component units comprised relaxation switching oscillators.[37]

The modelling studies described here have been largely based on experimental data obtained without external forcing to the biological system. This is possible because of the autorhythmic nature of gut electrical signals. In

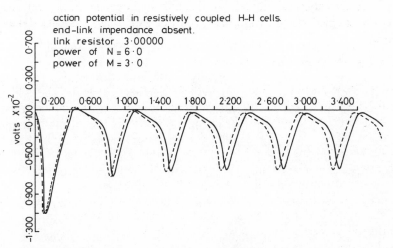

Fig. 7.10 *Two resistively coupled Hodgkin-Huxley type oscillators showing entrainment with small phase shift*

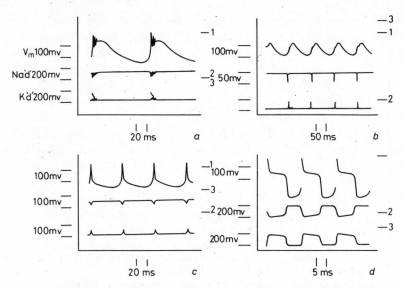

Fig. 7.11 *A family of wave-forms obtained from an electronic implementation of the Hodgkin-Huxley equations*

animals, some forcing experiments have been performed in which the gut is 'paced' with a regular pulse stimulus of current. In both the stomach and small intestine of dogs it has been possible to entrain the organ to the external pacing frequency. In the small intestine it was possible to reverse the direction of food transport using electrical pacing, showing the linkage between electrical 'slow waves' and mechanical muscular contractions. Pulse-synchronization studies have been made on the hardware oscillator models described above in an attempt to distinguish between the various oscillator dynamics as an acceptable hypothesis for the gut. It was found that the relaxation oscillator dynamic was unable to reproduce the physiological pacing experimental data.[38]

Although the gut model is inherently non-linear, considerable progress in the analysis of coupled oscillator arrays has been made recently. The principal mathematical tool used has been that of the Krylov-Bogolioubov method of 'harmonic balancing'. Using this technique the existence of multiple mode solutions in reactively coupled non-linear oscillators has been established. For parallel RLC coupling of Fig. 7.8 a parameter space has been established showing where multiple solutions are feasible,[39] as illustrated in Fig. 7.12. As well as having multiple solutions, coupled oscillator arrays can have the simultaneous existence of two individual modes, called a multimode. Analytical investigations of multimodes have been performed for both one- and two-dimensional arrays of oscillators.[40, 41] The multimode behaviour of a model involving the large-intestinal dynamic of eqn. (7.7) has also been analysed theoretically.[33] A considerable body of theoretical analysis is now

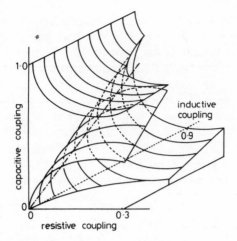

Fig. 7.12   *Boundary contours in RCL space separating stable and unstable regions for antiphase mode for coupled van der Pol oscillators*
For non-zero values of *L* there are two stable regions in the RC plane

available for the model structure of the digestive tract, although it must be emphasized that this is for small non-linearity with oscillator waveforms being almost sinusoidal.

This section has shown modelling studies of a biological system which is inherently non-linear and to which it is relatively difficult to apply forcing inputs. It would certainly be desirable to apply other pacing wave-forms such as sinusoids although this presents practical problems which currently have not been overcome. The model, however, has been successful in reproducing physiological data and sometimes in prediction of phenomena. At present, experimental data, modelling simulation and theoretical analysis appear to be keeping in step with one another. This is surely a healthy condition in that it helps to advance research in all directions with cross-fertilization of concepts between disciplines. An example of this is that spectral analysis of 'waxing and waning' phenomena in human duodenal signals has revealed the interesting possibility of multimodes with an indirect estimation of coupling parameters impossible from the conventional entrained single modes.[23]

7.5 Nerve impulse and conduction modelling

The development of the Hodgkin-Huxley equations for nerve fibre dynamics represents the most fertile example of mathematical modelling in physiology. The Nobel prize which they received in conjunction with J. C. Eccles demonstrates the importance which has been attached to this work. Their studies have initiated many research programmes into ionic structure of tissue and led countless mathematicians to pit their energies against these famous equations. The literature on the subject is vast and only a few points will be brought out in this section. It has been stated repeatedly that the combined physiological and mathematical studies performed by Hodgkin and Huxley represent a pattern of scientific investigation which any young researcher should read and seek to emulate. Their major mathematical contribution can be found in Reference 42. For a fascinating personal account of their joint studies the reader should consult their Nobel lectures.[43, 44]

Fitzhugh has studied an extensive range of mathematical models relating to nerve excitation and propagation, a broad treatment of which can be found in Reference 45. He remarks that earlier models of the nerve exist, such as that due to Young[46] and given by

$$\dot{V} = k_{11}(V - V_k) + k_{12}(U - U_k) + aI$$
$$\dot{U} = k_{21}(V - V_k) + k_{22}(U - U_k)$$

(7.8)

The two state variables in this model have been identified as V, the membrane potential, and U, an accommodation variable. The input variable is I, the stimulating current applied to the nerve through an electrode. This model has been extensively studied via classical phase-plane analysis. Fitzhugh suggested a generalization of the van der Pol equation, already given in eqn. (7.6), for nerve impulse studies. Such a formulation is consistent with an iron-wire model of a nerve axon[47] and the well-known tunnel-diode equivalent circuit.[48] Being second-order dynamics the Fitzhugh equations are readily amenable to phase-plane analysis.

In contrast to these simple models the Hodgkin-Huxley equations are fourth order and highly non-linear. The state variables are membrane potential V_m and three excitation parameters having the following equations.

$$-C_m \frac{dV_m}{dt} = g_{Na}(V_m + 115) + g_K(V_m - 12) + g_L(V_m + 10 \cdot 6)$$

$$\frac{dm}{dt} = \alpha_m + (\alpha_m + \beta_m)m \qquad 1 > m > 0$$

$$\frac{dn}{dt} = \alpha_n + (\alpha_n + \beta_n)n \qquad 1 > n > 0$$

$$\frac{dh}{dt} = \alpha_h + (\alpha_h + \beta_h)h \qquad 1 > h > 0$$

(7.9)

The ionic conductances are related by the following empirical equations

$$g_{Na} = 120m^3h \quad \text{m mho cm}^{-2}$$
$$g_K = 36n^4 \quad \text{m mho cm}^{-2}$$
$$g_L = 0 \cdot 3 \quad \text{m mho cm}^{-2}$$

(7.10)

The α and β terms are rate functions describing the time dependency on the membrane potential V_m of the ionic processes and are given by

$$\alpha_m = 0.1(25 + V_m)/(\exp{(0.1(V_m + 25))} - 1.0)$$

$$\alpha_n = 0.01(V_m + 10)/(\exp(0.1(V_m + 10)) - 1.0)$$

$$\alpha_h = 0.07(\exp{(0.05V_m)})$$

$$\beta_m = 4.0(\exp{(V_m/18.0)})$$

$$\beta_n = 0.125(\exp{(V_m/80.0)})$$

$$\beta_h = 1.0/(\exp{((V_m + 30.0)0.1)} + 1.0)$$

$$(7.11)$$

In equivalent circuit form these equations represent a parallel network comprising a linear membrane capacitance and three non-linear conductance branches due to sodium, potassium and leakage ionic pathways across the membrane.

When originally proposed Hodgkin-Huxley solved these equations by a numerical method using a desk calculator, and found good agreement with a wide range of experimental data. Digital computer simulations of these equations can be found in numerous places such as References 49 and 50.

To facilitate an approximate phase-plane analysis of the equations Fitzhugh reduced the system by arranging the four state variables, according to the order of magnitude of their relaxation times, into two classes. These are fast variables V_m and m, and slow variables h and n, having about a 10-to-1 ratio of time constants. In this way, instead of allowing h and n to vary slowly according to their differential equations, they can be assumed constant while V_m and m form a reduced system whose phase-plane can be analysed. In this way, numerous properties of the equations and their experimental counterparts can be eludicated.[51]

The differential eqns. (7.9) relate to a so-called 'space-clamped' nerve axon in which the membrane current and potential vary with time but not with distance. Under normal conditions the nerve impulse is conducted along the axon because of circulating currents which cross the membrane and flow lengthwise inside and outside the axon. In this case the axon should be represented by a ladder structure or cable which in the limit produces a partial differential equation formulation of the form

$$\frac{\partial V_m}{\partial t} = \frac{1}{C}\left[\frac{1}{\pi d(r_e + r_i)}\left(\frac{\partial^2 V}{\partial x^2} + r_e J_a\right) - I_i\right]$$

$$\frac{\partial W_j}{\partial t} = F_j(V, W_1, \ldots, W_\mu)$$

$$(7.12)$$

where r_i and r_e are internal and external resistances per unit length for the membrane, d is the diameter of the axon, and J_a is the stimulating current density. The numerical solution of the distributed form of the Hodgkin-Huxley equations has been considered by many people. For example, Lieberstein has studied the effect of inductance in the model,[52] while Bellman and co-workers have applied the method of differential quadrature to their simulation.[53] In this latter method the partial differential equation is reduced to a set of ordinary differential equations using quadrature-weighted coefficients obtained via cubic-spline methods.

In a similar way to which the complete Hodgkin-Huxley equations have been reduced to facilitate analysis and simulations, a range of electronic models have been proposed for studying nerve impulses. A very complex model was that due to Lewis,[54] while simulation of variable conductances using field-effect transistors forms the basis of studies of Gulrajani and Roberge[55] and a simplified version by Roy.[56] It is not surprising that alternative dynamics to the Hodgkin-Huxley equations have been hypothesized for nerve-action potentials. An example is the work of Zeeman[57] who used catastrophe theory to analyse his somewhat simpler formulation. As an example of the degree of fit obtained between experimental data and the model Fig. 7.13 shows the original results of Hodgkin and Huxley. A comparison between the simplified Zeeman dynamic and the full Hodgkin-Huxley equations is given in Fig. 7.14 for time cause changes in conductance paths for a propagated action potential.

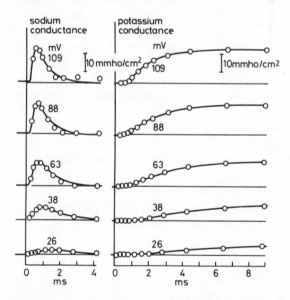

Fig. 7.13 *Time course changes in sodium and potassium conductance in the space-clamped squid axon*
Circles indicate experimental data, smooth curves indicate Hodgkin-Huxley equation responses
(From Hodgkin, A. L., and Huxley, A. F. *J. Physiol., 1952,* **117,** pp. 500–544)

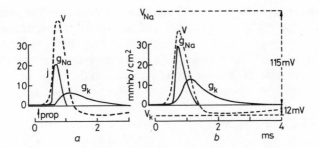

Fig. 7.14 *Propagated action potential and conductances for (a) Zeeman model (b) Hodgkin-Huxley model*
(From Zeeman, E. C., 'Differential equations for the heartbeat and nerve impulse., in 'Towards a Theoretical Biology 4', Edinburgh University Press, 1972)

The Hodgkin-Huxley equations have formed the basis for related mathematical modelling studies in many other tissue structures. In these cases physiological measurements are usually much more difficult than it was for the giant axon of the squid, where the diameter of a single nerve fibre could be up to 1 mm. In particular, Noble produced an equivalent modified version of the Hodgkin-Huxley equations to describe the dynamics of Purkinje fibre action and pacemaker potentials.[58] As already mentioned in Section 7.4 these dynamics have been further modified to make them applicable to gastrointestinal electrical potential generation. It is clear that non-linear dynamics of this type have spawned many avenues of research, linking mathematicians and physiologists in a unique fashion. In equations of such complexity the problem of parameter estimation becomes almost insurmountable in most cases, and model validation is a major hurdle to overcome.

7.6 Modelling the glucoregulatory system

One of the most important regulation systems in the body is that of glucose homeostasis. Glucose is the prime energy supply to the muscles and the brain. Glucose exists explicitly in the blood, and implicitly in the form of aminoacids, glycogen, free fatty acids (FFA) and triglycerides in the blood, muscle and adipose tissue, and the liver. Glucose is formed by the process called gluconeogenesis, with excess glucose being converted into its implicit forms by processes which include enhancement by pancreatic release of insulin. Conversely, deficiencies in glucose are offset by glycogenolysis which includes the inverse processes to gluconeogenesis. The control of these interacting processes forms the glucoregulatory system.

Since 1960 there have been a number of attempts at representing the glucose metabolism and its control in mathematical terms. These models have been of the lumped parameter and compartmental type, although certain entirely conceptual models also have been formulated. The mathematical models fall into categories of linear and non-linear formulations, with the linear attempts appearing first historically. The early linear models of Bolie[59] and Ackerman *et al.*[60] used organ-level processes only. The Ackerman model had two compartments: one for blood glucose and the other for blood insulin interacting through a feedback loop, and represented by two linear equations

$$\frac{dg}{dt} = -m_1 g - m_2 h + J(t)$$

$$\frac{dh}{dt} = -m_3 h + m_4 g + K(t)$$

(7.13)

where g and h represents glucose- and insulin-level deviations, J and K are experimental input rates of infusion of glucose and insulin, and m_1, m_2, m_3 and m_4 are parameters. Equations (7.13) were reduced to a single second-order differential equation with the 'normal' natural frequency obtained from glucose tolerance tests. An attempt was then made to correlate this 'normal' parameter with diabetic cases. Not surprisingly for biological systems with large variations between normal subjects such a crude diagnostic method did not prove successful.

More recent models have reflected the uniform trend in biological modelling to non-linear systems. Typical of these much more complex models are those due to Foster,[61] Srinivasan *et al.*,[62] and Cramp and Carson.[63] In these models intermediary metabolic processes are included along with organ level processes. Foster's model was a two-substrate (glucose and FFA) and two-hormone (insulin and glucagon) structure. Its non-linearities were produced via function generators giving piece-wise linear curves. The Srinivasan model was a two-substrate (glucose and FFA) and four-hormone (insulin, glucagon, epinephrine and growth hormone) structure. Non-linearities such as threshold and saturation levels were generated with a hyperbolic tangent function which had the following characteristics: its lower and upper saturation levels and the slope were adjustable via three parameters, while the upper saturation value, when postulated to be parameter dependent, was linearly dependent. This was an extremely complex model with about forty parameters in the glucose portion alone.

The Carson model attempts to go beyond the somewhat arbitrary nature of the non-linearities in the Foster and Srinivasan models by including unit biochemical processes for all the metabolic pathways that they considered. The block diagram for their model is shown in Fig. 7.15. The individual equations for each pathway labelled in this diagram are non-linear by virtue of quotient terms and threshold characteristics.

Parameter estimation and model validation are of crucial importance in all of these models. The Foster model was validated with oral and intravenuous glucose injection (usually of an impulse nature), and studies were made to

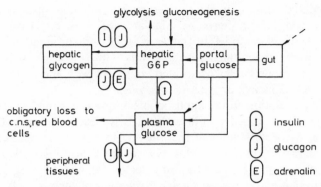

Fig. 7.15 *Compartmental model of liver glucose metabolism and its hormonal control*
(From Cramp, D. G., and Carson, E. R., 'The Dynamics of Blood Glucose and its
Regulating Hormones', *in* 'Biological Systems, Modelling and Control., Peregrinus,
1979).

characterize the diabetic state. His findings were that the normal glucose
tolerance test was dominated by changes in insulin secretion dependent on
changes in the level of glucose. The Srinivasan model was validated with
about two hour long observations obtained from three-minute intravenous
pulsed glucose or insulin injections. Responses of plasma glucose, insulin and
FFA were fitted by the model to experimental data. An example of the model
fit is shown in Fig. 7.16. Further simulations were performed on the model

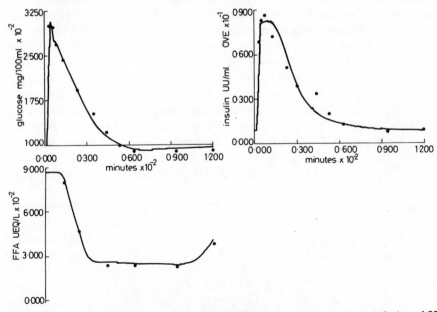

Fig. 7.16 *Experimental and Srinivasan model results for intravenous glucose infusion of 35
gm at a constant rate over 3.5 minutes*

using pulsed and continuous infusion (equivalent to a step input) of glucagon, epinephrine and growth hormone. Similar validation tests have been performed on the Carson model. In this case oscillations occurred in the gluco-regulatory system under glucose loading tests. This is in contrast to the other two non-linear models referred to, and consistent with spontaneous oscillations observed in fasting, unanaesthetized dogs.[64]

The steady evolution from simplistic, black-box linear models to complex, unit physiologically based non-linear models is clearly demonstrated in this summary of dynamic studies of the human glucose regulation system. The complex interacting nature of this system is further enlarged when it is realized that many other contributing factors should be included in the already daunting mathematical model. Validation of such models with their attempts to keep close to the biochemistry of the physiological processes becomes an urgent but difficult matter.

7.7 Modelling and analysis of a biochemical loop

This brief section is included to illustrate biological dynamics study and analysis at the microscopic- rather than the large-organ level. The particular system considered is that of glycolytic oscillations which have been experimentally observed via the fluorescent property of NADH (nicotinamide-adenine denucleotide). This enables fluorometry to be used in measuring the conditions of a certain yeast extract.[65] Basically there have been two types of model studied, one comprising two components with complex interaction, and the other comprising many components with a single feedback connection. These structures are shown in Fig. 7.17.

For the two-components model a linearized version has been used to give the necessary conditions for oscillation.[66] In the studies by Hess and his co-workers the two components were taken to be substrate concentration α of ATP (adenosine triphosphate) and product concentration γ of ADP (adenosine diphosphate). The model equations used were

$$\frac{d\alpha}{dt} = \sigma_1 - \sigma_M \Phi \qquad \frac{d\gamma}{dt} = \sigma_M \Phi - K_8 \gamma \qquad (7.14)$$

where

$$\Phi = \frac{\left(\dfrac{\alpha}{\epsilon+1}\right)\left(1 + \dfrac{\alpha}{\epsilon+1}\right)(1+\gamma)^2 + L\theta\left(\dfrac{\alpha C}{\epsilon'+1}\right)\left(1 + \dfrac{\alpha c}{\epsilon'+1}\right)}{L\left(1 + \dfrac{\alpha c}{\epsilon'+1}\right)^2 + (1+\gamma)^2\left(1 + \dfrac{\alpha}{\epsilon+1}\right)^2}$$

The meaning of the various model parameters, together with typical values, can be found in Reference 65. The presence of L indicates a so-called 'allosteric'

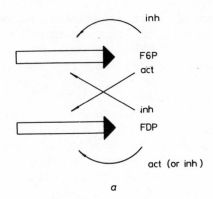

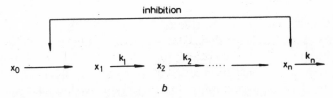

Fig. 7.17   *Block diagram of two model structures for biochemical oscillations*
         (a) Two-component, cross-coupled system
         (b) Chain-reaction structure with end-product inhibition (feedback)

reaction having sigmoidal kinetics. For small values of L the model reduces to a Michaelian reaction having hyperbolic kinetics. Stability analysis of these two types of kinetics has shown that the Michaelis system produces oscillations only for an extremely small range of substrate input rate.[67] In contrast, the allosteric (or Monod model) system produces periodic oscillations for both steady and periodic perturbations. It also gives a wide range of entrainment phenomena. In experimental studies Hess has reported entrainment of the glycolytic reaction to a periodic rate of substrate injection over a period range $0.7 \leq T/T_0 \leq 1.2$. The model showed a similar entrainment region $(0.89 \leq T/T_0 \leq 1.11)$, together with subharmonic entrainment at ratios of $1/2$ and $1/3$. Simplified versions of this model structure have been studied by Pavlidis[68] and studied via phase-plane analysis and phase-response curves used in circadian rhythm research.

The second type of model comprises a chain of chemical sequences with end-product inhibition in a single feedback loop. The substrate concentration x_0 is considered to be constant, and the first reaction is controlled by the end-product concentration according to the rate law

$$h(x_n) = \frac{K}{1 + \alpha x_n^\rho} \tag{7.15}$$

where ρ is the Hill coefficient for the inhibition. The equations typically considered are given by

$$\frac{dx_1}{dt} = h(x_n) - b_1 x_1$$

$$\frac{dx_j}{dt} = g_{j-1} x_{j-1} - b_j x_j \qquad j = 2, \ldots, n$$

(7.16)

Sel'kov[69] recognized that this system could have an unstable singularity for certain values of ρ and n, while Goodwin[70] demonstrated limit-cycle oscillations for $n = 3$ and $\rho = 1$ by computer simulations. Higgins[71] showed that the singularity is stable unless

$$\rho > \frac{1}{(\cos \pi/n)^n}$$

The same equations have been studied by a number of other workers using Liapunov theory[72] and the describing-function technique.[73] Using the latter method, Rapp considered a d.c. term together with a fundamental component for the system under oscillation and then employed the dual-input-describing function to determine conditions for periodic behaviour. He showed that when $\rho = 1$, no oscillations are possible for any value of n. For $\rho = 2$, stable oscillations can occur for $n > 8$ and, for $\rho = 3$ and $\rho = 4$, oscillations can occur for $n > 5$. The method can be extended to include delays in the linear dynamics, which have often been considered to be present in metabolic loops. Rapp has considered other extensions such as multiple non-linearities separated by linear components, and parallel feedback paths. A further refinement is the possibility that biochemical control loops can sometimes oscillate in the presence of randomly varying inputs,[65] for which the describing-function method can be adapted.[74]

In the dynamic modelling of the glycolytic control loop one has the possibility of closely controlled experimental studies which have included the standard systems forcing functions of steps, sinusoids and random sequences. Perhaps more importantly, individual biochemical processes can be specified for further detailed exploration of dynamic behaviour. Once again, the model is non-linear but some analytical progress has been made in prediction of mathematical performance. Such studies are all too rare in biological modelling research.

7.8 Comments

The examples of biological modelling described briefly in the foregoing sections reflect in part the personal interests of the author. There are many areas which have not been considered at all, but it is considered that the systems which have been described illustrate the major features common in this inter-disciplinary research. For a collection of papers relating to other fields the reader can consult the Proceedings of the IFAC Conference on Regulation and Control in Physiological Systems.[75] Other biological dynamics include pharmokinetics, thermoregulatory systems, muscle and limb dynamics, the whole cardiovascular system including ECG generation and blood fluid flow characteristics, neuron populations and the CNS (Central Nervous System).

A number of factors stand out in these attempts at modelling. Firstly, initial models usually comprise linear dynamics of low order. These inevitably prove to be invalid for physiologically meaningful operating ranges. They do, however, point the way to more realistic dynamics, and of course have the advantage that they can be analysed using control systems theory. Secondly, the conventional forcing functions used in engineering systems identification can be applied successfully to biomedical structures particularly if linear models are a reasonable approximation. Thus, several examples have been cited of the use of impulses, steps, sinusoids and PRBS input functions. In certain cases, parameter estimation via least squares or maximum likelihood has been performed satisfactorily, but at this point the third factor becomes important. This is the chronic non-linearity of the dynamic biological world. Maintenance of physiological relevance and insight into the biochemical processes always requires non-linearities to be recognized. Under these condi-tions the analytical tools at the systems analyst's disposal become significantly fewer, less robust and less useful. The fourth factor is that oscillations are a common feature in biological systems. Unlike engineering applications, oscil-lating loops appear to be both normal and desirable, rather than unusual and disastrous. Recognizing this feature alleviates in small measure the analytical problems. For example, the describing-function technique has a major con-tribution in being able to predict the amplitude, frequency and stability of limit-cycle oscillations. Also, the Krylov-Bogolioubov method in matrix form is well suited to high-order interacting oscillatory structures. Further, har-monic balancing methods give algebraic methods of solution for quite general structures under oscillating conditions.

In conclusion it is observed that although biological modelling constantly faces the problem of non-linear dynamics, the analysis of such systems is advancing the theoretical tools which will enable us to gain further insight into more complex interacting structures.

7.9 References

1 LINKENS, D. A.: 'Biological Systems, Modelling and Control' (Peter Peregrinus, 1979)
2 STARK, L.: 'Stability, oscillations, and noise in the human pupil servomechanism', 1959, *Proc. IRE*, **47**, pp. 1925–1936
3 STARK, L., and SHERMAN, P. M.: 'A servoanalytic study of the consensual pupil reflex to light', *J. Neurophysiol.*, 1957, **20**, pp. 17–26

4 STERN, H. J.: 'Simple method for early diagnosis of abnormalities of pupillary reaction', *Br. J. Opthalmol.*, 1944, **28**, pp. 276–278

5. STARK, L.: 'Biological rhythms, noise, and asymmetry in the pupil-retinal control system', *Ann. N.Y. Acad. Sci.*, 1962, **98**, pp. 1096–1108

6 CLYNES, M.: 'Unidirectional rate sensitivity. A biocybernetic law of reflex and humoral systems as physiological channels of control and communication', *Ann. N.Y. Acad. Sci.*, 1961, **93**, pp. 946–969

7 STARK, L.: 'Neurological control system. Studies in bioengineering' (Plenum Press, New York, 1968)

8 GRODINS, F. S., GRAY, J. S., SCHROEDER, K. R., NORRIS, A. L., and JONES, R. W.: 'Respiratory responses to CO_2 inhalation. A theoretical study of a nonlinear biological regulator', *J. Appl. Physiol.*, 1954, **7**, pp. 283–308

9 GRODINS, F. S., BUELL, J., and BART, A. J.: 'Mathematical analysis and digital simulation of the respiratory control system', *J. Appl. Physiol.*, 1967, **22**, pp. 260–276

10 SWANSON, G. D., and BELLVILLE, J. W.: 'Step changes in end-tidal CO_2: methods and implications', *J. Appl. Physiol.*, 1975, **39**, pp. 377–385

11 MICHELS, D. B., WARD, D. S., AQLEH, K. A., SWANSON, G. D., and BELLVILLE, J. W.: 'End-tidal CO_2 $-O_2$ response to voluntary ventilation manoeuvres', *Proc. San Diego Biomed. Symp.*, 1974, **13**, pp. 259–266

12 CASABURI, R., WHIPP, B. J., WASSERMAN, K., BEAVER, W. L., and KOYAL, S. N.: 'Ventilatory and gas exchange dynamics in response to sinusoidal work', *J. Appl. Physiol.*, 1977, **42**, pp. 300–311

13 FUJIHARA, Y., HILDEBRANDT, J., and HILDEBRANDT, J. R.: 'Cardiorespiratory transients in exercising man—II. Linear models', *J. Appl. Physiol.*, 1973, **35**, pp. 68–76

14 SWANSON, G. D., and BELLVILLE, J. W.: 'Dynamic end-tidal forcing for respiratory controller parameter estimation', *in* 'Regulation and Control in Physiological Systems', Iberall and Guyton (Eds.), Pittsburgh, PA, Instrument Society of America, 1973, pp. 269–273

15 WHIPP, B. J., CASABURI, R., and WASSERMAN, K.: 'Determinants of the dynamics of ventilation during muscular exercise', Symp. 'Modelling of a Biological Control System: the Regulation of Breathing', Sept., 1978, Oxford, p. 174

16 BENNETT, F. M., REISCHL, P. R., GRODINS, F. S., YAMASHIRO, S. M., and FORDYCE, W. E.: 'Identification of the dynamics of the human ventilatory response to exercise', Symp. 'Modelling of a Biological Control System: the Regulation of Breathing', Sept., 1978, Oxford, p. 173

17 SWANSON, G. D.: 'Biological signal conditioning for system identification', *Proc. IEEE*, 1977, **65**, pp. 735–740

18. SWANSON, G. D.: 'Input stimulus design for model discrimination in human respiratory control', Symp. 'Modelling of a Biological Control System: the Regulation of Breathing', Sept., 1978, Oxford, p. 165

19. WIBERG, D. M., BELLVILLE, J. W., BROVKO, O., MAINE, R., and TAI, T. C.: 'Modelling and parameter identification of the human respiratory system', *Proc. IEEE Conf. on Decision and Control*, Jan., 1979, San Diego

20 SWANSON, G. D.: 'Evaluation of the Grodins respiratory model via dynamic end-tidal forcing', *Am. J. Physiol.*, 1977, **233**, pp. 66–72

21 CHRISTENSEN, J., SCHEDL, H. P., and CLIFTON, J. A.: 'The small intestine basic electrical rhythm (slow wave) frequency gradient in normal man and in patients with a variety of diseases', *Gastroenterology*, 1966, **50**(3), pp. 309–315

22 TAYLOR, I., DUTHIE, H. L., SMALLWOOD, R. H., and LINKENS, D. A.: 'Large bowel myoelectrical activity in man', *GUT*, 1975, **16**, pp. 808–814

23 SMALLWOOD, R. H., LINKENS, D. A., and STODDARD, C. J.: 'Amplitude fluctuations in human duodenal slow-waves: computer analysis and modelling implications', 7th Int. Symp. on 'Gastrointestinal Motility', Sept., 1979, Iowa

24 SZURZWESKI, J. H.: 'A migrating electric complex of the canine small-intestine', *Am. J. Physiol.,* 1969, **217,** pp. 1757–1763

25 VANTRAPPEN, G., JANSSENS, J., HELLEMANS, J., and GHOOS, Y.: 'The interdigestive myoelectric complex in normal subjects and patients with bacterial overgrowth in the jejunum', 6th Int. Symp. on Gastrointestinal Motility, Sept. 1977, Edinburgh

26 DUTHIE, H. L.: 'Electrical activity of gastrointestinal smooth muscle', *GUT,* 1974, **15,** pp. 669–681

27 NELSEN, T. S., and BECKER, J. C.: 'Simulation of the electrical and mechanical gradient of the small-intestine', *Am. J. Physiol.,* 1968, **214,** pp. 749–757

28 SARNA, S. K., DANIEL, E. E., and KINGMA, Y. J.: 'Simulation of slow-wave electrical activity of small-intestine', *Am. J. Physiol.,* 1971, **221,** pp. 166–175

29 ROBERTSON-DUNN, B., and LINKENS, D. A.: 'A mathematical model of the slow-wave electrical activity of the human small intestine,' *Med. and Biol. Eng.,* 1974, **12**(6), pp. 750–755

30 LINKENS, D. A., and MHONE, P. G.: 'Frequency transients in a coupled oscillator model of intestinal myoelectrical activity', *Comp. in Biol. and Med.,* 1979, **9,** pp. 131–143

31 SHEARIN, N. L., BOWES, K. L., KINGMA, Y. J., and KOLES, Z. J.: 'Frequency analysis of electrical activity in dog colon', 6th Int. Symp. on 'Gastrointestinal Motility', Sept., 1977, Edinburgh

32 LINKENS, D. A., and DATARDINA, S. P.: 'Human colonic modelling and multiple solutions in non-linear oscillators', 6th Int. Symp. on 'Gastrointestinal Motility', Sept., 1977, Edinburgh

33 DATARDINA, S. P., and LINKENS, D. A.: 'Multimode oscillations in mutually coupled van der Pol type oscillators with fifth power non-linear characteristics', *IEEE Trans. Circuits & Syst.,* **CAS-26,** 1978, pp. 308–315

34 LINKENS, D. A., and DATARDINA, S. P.: 'Frequency entrainment of coupled Hodgkin-Huxley type oscillators for modelling gastro-intestinal electrical activity', *IEEE Trans. Biomed. Eng.,* 1977, **BME24**(4), pp. 362–365

35 PATTON, R. J., and LINKENS, D. A.: 'Phenomenological investigation of a distributed parameter model for co-ordinating the mechanical activity of the mammalian gut', IFAC 2nd Symp. on 'Control of Distributed Parameter Systems', June, 1977, Univ. of Warwick, pp. 1–17

36 PATTON, R. J., and LINKENS, D. A.: 'Hodgkin-Huxley type electronic modelling of gastro-intestinal electrical activity', *Med & Biol. Eng. & Comput.,* 1978, **16,** pp. 195–202

37 BROWN, B. H., DUTHIE, H. L., HORN, A. R., and SMALLWOOD, R. H.: 'A linked oscillator model of electrical activity of human small-intestine', *Am. J. Physiol.,* 1975, **229,** pp. 384–388

38. LINKENS, D. A.: 'Canine colonic pacing and coupled oscillator synchronization', *J. Physiol.,* March, 1978, **276,** p. 37P

39 LINKENS, D. A.: 'The stability of entrainment conditions for RLC coupled van der Pol oscillators', *Bull. Math. Biol.,* 1977, **39,** pp. 359–372

40 ENDO, T., and MORI, S.: 'Mode analysis of a multimode ladder oscillator', *IEEE Trans. & Circuits & Syst.,* 1976, 23, pp. 100–113

41 ENDO, T., and MORI, S.: 'Mode analysis of a two-dimensional low-pass multimode oscillator', *IEEE Trans. Circuits & Syst.,* 1976, **23,** pp. 517–530

42 HODGKIN, A. L., and HUXLEY, A. F.: 'A quantitative description of membrane current and its application to conduction and excitation in nerve', *J. Physiol.,* 1952, **117,** pp. 500–544

43 HODGKIN, A. L.: 'The ionic basis of nervous conduction', *Science,* Sept., 1964, **145,** pp. 1148–1153

44 HUXLEY, A. F.: 'Excitation and conduction in nerve: quantitative analysis', *Science,* Sept., 1964, pp. 1154–1159

45 FITZHUGH, R.: 'Mathematical models of excitation and propagation in nerve', *in* 'Biological Engineering', SCHWAN, H. P. (Ed) (McGraw-Hill, 1969)

46 YOUNG, G.: 'On reinforcement and interference between stimuli', *Bull. Math. Biophys.*, 1941, **3**, pp. 5-12

47 BONHOEFFER, K. F.: Activation of passive iron as a model for the excitation of nerve', *J. Gen. Physiol.*, 1948, **32**, pp. 69-91

48 NAGUMO, J., ARIMOTO, S., and YOSHIZAWA, S.: 'An active pulse transmission line simulating nerve axon', *Proc. IRE*, 1962, **50**, pp. 2061-2070

49 COLE, K. S., ANTOSIEWICZ, H. A., and RABINOWITZ, P.: 'Automatic computation of nerve excitation', *J. Soc. Ind. Appl. Math.*, 1955, **3**, 3

50 COOLEY, J. W., and DODGE, F. A.: 'Digital computer solutions for excitation and propagation of the nerve impulse', *Biophys. J.*, 1966, **6**, pp. 583-597

51 FITZHUGH, R.: 'Impulses and physiological states in nerve models', *Biophys. J.*, 1961, **1**, pp. 446-466

52 LIEBERSTEIN, H. M.: 'On the Hodgkin-Huxley partial differential equation', *Math. Biosci.*, 1967, **1**, p. 45

53 BELLMAN, R., KASHEF, B. G., and CASTI, J.: 'Differential quadrature; a technique for the rapid solution of non-linear partial differential equations', *J. Comp. Phys.*, 1972, **10**, p. 40

54 LEWIS, E. R.: 'An electronic model of neuroelectric point processes', *Kybernetik*, 1968, **5**, 1, pp. 30-46

55 GULRAJANI, R. M., and ROBERGE, F. A.: 'The modelling of the Hodgkin-Huxley membrane with field effect transistors', *Med. and Biol. Eng.*, Jan., 1976, pp. 31-40

56 ROY, G.: 'A simple electronic analog of the squid axon membrane', *IEEE Trans. Biomed. Eng.*, Jan., 1972, pp. 60-63

57 ZEEMAN, E. C.: 'Differential equations for the heartbeat and nerve impulse', *in* 'Towards a Theoretical Biology 4' (Edinburgh University Press, 1972)

58 NOBLE, D.: 'A modification of the Hodgkin-Huxley equations applicable to Purkinje fibre action and pace-maker potentials', *J. Physiol.*, 1962, **160**, pp. 317-352

59 BOLIE, V. W.: 'Coefficients of normal blood glucose regulation', *J. Clin. Inv.*, 1960, **39**, pp. 783-788

60 ACKERMAN, E., ROSEVEAR, J. W., and McGUKIN, W. F.: 'A mathematical model of the glucose tolerance test', *Phys. Med. & Biol.*, 1964, **9**, pp. 203-213

61 FOSTER, R. O.: 'Computer simulation of the glucose regulatory system in man', *Diabetes*, 1970, **19**, pp. 373

62 SRINIVASAN, R., KADISH, A., and SRIDHAR, R.: 'A mathematical model for the control mechanism of FFA-Glucose metabolism in normal humans', *Comput. & Biomed. Res.*, 1970, **3**, pp. 146-166

63 CRAMP, D. G., and CARSON, E. R.: 'The dynamics of blood glucose and its regulating hormones', *in* 'Biological Systems, Modelling and Control', Chap. 5. (Peter Peregrinus, England, 1979)

64 YATES, F. E., MARSH, D. J., SMITH, S. W., OOKHTENS, M., and BERGMAN, R.: 'Modelling metabolic systems and the attendant data handling problems', *in* IFAC Conf. on 'Regulation and Control in Physiological Systems', Iberall & Guyton (Eds.), pp. 464-469 (Rochester, New York, 1973)

65 HESS, B.: 'Oscillations in biochemical and biological systems', *Bull. IMA*, 1976, pp. 6-10

66 HIGGINS, J.: 'The theory of oscillating reactions', *Ind. Eng. Chem.*, 1967, **59**, pp. 18-62

67 SEL'KOV, E. E.: 'Self-oscillations in glycolysis—1. A simple kinetic model', *Eur. J. Biochem.*, 1968, **4**, pp. 79-86

68 PAVLIDIS, T.: 'Biological Oscillators: Their Mathematical Analysis' (Academic Press, New York, 1973)

69 SEL'KOV, E. E.: 'Scientific Thought', Kiev, 1965

70 GOODWIN, B. C.: 'Oscillatory behaviour in enzymatic control processes', *in* WEBER, G. (Ed.), 'Advances in Enzyme Regulation', 1965, 3, pp. 425-438 (Academic Press, New York, 1965)

71 HIGGINS, J., FRENKEL, R. HULME, E., LUCAS, A., and RANGAZAS, G.: 'The control theoretic approach to the analysis of glycolytic oscillators', *in Chance*, 1973, pp. 127–176

72 WALTER, C. F.: 'The occurrence and the significance of limit-cycle behaviour in controlled biochemical systems', *J. Theor. Biol.*, 1970, **27**, pp. 259–272

73 RAPP, P.: 'Mathematical techniques for the study of oscillations in biochemical control loops', *Bull. IMA*, Jan., 1976, pp. 11–21

74 GELB, A., and VAN DER VELDE, W.: 'Multiple-input Describing Functions and Nonlinear System Design', (McGraw-Hill Book Co., USA, 1968)

75 IBERALL, A. S., and GUYTON, A. C. (Eds.): 'Regulation and control in physiological systems', IFAC Symp., Rochester, New York, Pittsburgh, Pennsylvania, Inst. Soc. of America, 1976

Index